T0350893

Diversity and Inclusion in Industry

Today, more than ever, diversity, equity, and inclusion (DE&I) play a crucial role in organizational success, especially in industry, a sector that is sometimes overlooked. This book brings a new perspective on the implementation of diversity and inclusion in industry, including engineering, construction, manufacturing, etc. Data-driven longitudinal studies show the positive economic impact of diversity in these fields. A strong array of case studies is included, and the authors' firsthand experiences provide information to industry professionals to help them understand the success and benefits that diversity can bring to these fields and how to embrace them outside of a corporate setting.

FEATURES

- Discusses the DE&I role in the industry sector specifically
- Includes numerous case studies from industry giants and small companies
- Explains layers of diversity in line management
- Shows the correlation of diversity to the prosperity of companies over time
- Identifies diversity as an important tool for future growth

This book is intended for professionals as well as students in upper-level undergraduate or graduate programs that are interested in or currently studying workplace diversity.

Diversity and Inclusion in Industry
A Road to Prosperity

Rengasamy Kasinathan
Munikiran Mallu
Madalyn Bozinski

CRC Press
Taylor & Francis Group
Boca Raton London New York

CRC Press is an imprint of the
Taylor & Francis Group, an **informa** business

First edition published 2024
by CRC Press
2385 NW Executive Center Drive, Suite 320, Boca Raton FL 33431

and by CRC Press
4 Park Square, Milton Park, Abingdon, Oxon, OX14 4RN

CRC Press is an imprint of Taylor & Francis Group, LLC

British Library Cataloguing-in-Publication Data
A catalogue record for this book is available from the British Library

ISBN: 978-1-032-37341-6 (hbk)
ISBN: 978-1-032-37592-2 (pbk)
ISBN: 978-1-003-34096-6 (ebk)

DOI: 10.1201/9781003340966

Typeset in Minion
by MPS Limited, Dehradun

To my late parents (Mr. Rengasamy and Mrs. Thengani Ammal),
my wife (Dr. Sumathi Kasinathan), my daughter
(Dr. Sushma Kasinathan) and son (Shriram Kasinathan)
– Rengasamy Kasinathan

To my son (Vishruth Reddy Mallu) and daughter
(Sreshta Reddy Mallu)
– Munikiran Mallu

To my parents (Glenn and Laura Bozinski) and my partner
(Iain Wright)
– Madalyn Bozinski

Contents

Preface

This book is intended to provide insights and guidance on how diversity and inclusion can bring prosperity to business, specifically in Industry. In this case, Industry is defined as a sector of business comprised of technical work such as science, technology, engineering, and mathematics (STEM), manufacturing, construction, and other related fields. This text contains sufficient material to be used as a guide for Industry professionals looking to improve diversity and inclusion at their company. This material is also suitable for master's level students of any background but is tailored to those familiar with Industry.

Diversity and Inclusion have been at the forefront of cultural conversation for several decades. Its relevance and importance in the workplace only continue to increase as society advances. It is widely known that Industry has historically fallen behind diversity progress in the workplace, as it has traditionally been lacking minority representation. This book addresses how diversity and inclusion can be addressed in Industry and looks at several cases of diversity and inclusion initiatives in this sector.

The book is structured to create a comprehensive understanding of why diversity is necessary, strategies for implementing diversity and inclusion in Industry, and how these efforts can ultimately lead to prosperity.

Chapter 1 details the need for diversity in the workplace broadly as well as in industry. As the makeup of the U.S. population changes, companies are seeing the need to alter business and hiring practices from the way things have been run for decades or even centuries. A diverse workforce fosters innovation and creativity. Having workers from a wide array of backgrounds allows for vastly different experiences and knowledge to meld together, allowing for innovation and new ideas that would never have been fostered in a non-diverse group. Since its beginning, Industry has been dominated predominantly by white males, and most of the industry workforces continues to be made up of that demographic. Lack of gender diversity is one of the major critiques of Industry, specifically STEM fields such as engineering, as well as other fields like construction. Women are underrepresented in these fields, and they also face a significant gender wage gap. This lack of diversity can hinder innovation, a key asset needed for most successful companies in Industry.

Chapter 2 delves deeper into how diversity and inclusion can be managed at the corporate level, or "from the top". Ensuring diversity at a corporate level does not equate to simply enacting affirmative action policies (AAP). AAP are procedures that can be implemented to consciously advance qualified minorities in the workplace. Managing diversity includes understanding a broad depth of insights and experiences that those at the topmost level of a company can utilize to bring people to their teams who sometimes may be overlooked. This can be done through actions such as placing values on unconventional skill sets, or broadening the idea of what proper qualifications for a position could look like.

Chapter 3 discusses how diversity can be implemented throughout a company by examining current trends in some of the most diverse businesses. It also discusses challenges faced by companies that have struggled to achieve diversity in some or all forms. As with any new endeavor, implementing diversity initiatives in the workplace can present challenges. The workplaces and employees' standards of what is expected from their employers are ever-growing and evolving. Companies must be well-equipped for the challenges they may face and understand the strategy to overcome them.

Chapter 4 looks into diversity implementation at the management level and the importance of the role that management plays in successful diversity initiatives. Implementation of diversity and inclusion initiatives is a critical task for employees at all levels of management. Each level of management is responsible for different aspects of the success of an initiative. While executives and upper management are responsible for creating the diversity program and taking high-level actions for its conception, mid-level managers are the individuals who can have the largest impact on whether these programs succeed.

Chapter 5 consequently looks at the role of the executive level and C-Suite in diversity implementation. Research has shown that a company's diversity policy will not be effective unless every level of the company is engaged in the mission of diversity and inclusion. This is especially true of a company's C-Suite. These executives are responsible for setting the mission and high-level goals of the company. Without their active support and engagement in these initiatives, it is unlikely that a strong inclusive culture will be able to take off at other levels throughout the company.

Chapter 6 discusses the economic benefits of diversity. Diversity drives economic benefits for companies. A diverse workforce is essential to a strong economy. Businesses that embrace diversity have a more solid footing in the marketplace than others. A diverse workforce combines workers from different backgrounds and experiences that together breed a more creative, innovative, and productive workforce. In addition, businesses have learned that they can draw upon our nation's diversity to strengthen their bottom line. In this way, diversity is a key ingredient to growing a strong and inclusive economy that is built to last.

Chapter 7 details how a successful diversity and inclusion strategy can be created and implemented. Proper implementation of diversity and inclusion initiatives in the workplace can be challenging. It is important that companies have an implementation plan that they can stick to ensure that their initiatives are successful. Diversity managers should be cognizant of the common reasons why these initiatives fail so they can avoid them or correct these mistakes. It is also important for companies to study what diversity programs have had legitimate success in the past and work to implement those instead of common strategies that have been proven to fail.

Chapter 8 examines several case studies of both large and small engineering companies that have had either successes or struggles with diversity programs. Every company can have challenges and triumphs when it comes to diversity programs. Fields such as science, engineering, and technology have especially unique challenges when it comes to issues such as gender diversity and diverse recruiting. Some companies have found unique and successful solutions for addressing these problems. While there is a set of strategies that have often been successful, it is important for a company to be introspective and to carefully examine how diversity can really thrive.

The book concludes with Chapter 9, which takes a look at the lessons learned throughout the previous chapters. It also looks forward at modern challenges, specifically the COVID-19 pandemic, which can alter the landscape of diversity in the workplace.

Authors

Rengasamy Kasinathan, PhD, MBA, is the founder and the president of Environmental Engineering Solutions, PC, a consulting engineering firm located in Westchester, New York. He earned a doctoral degree at SUNY Syracuse in ESF bio process engineering, focusing on conversion of paper mill fines to sugar and ultimately to bioplastics, earned an MBA at Fayetteville State University of the University of North Carolina, an MS in energy and atmospheric science at New York University, and a BS in chemical engineering at Annamalai University in India. He is a certified air quality modeler and has performed several air quality studies and secured many permits for hospitals, universities, and other industries throughout the United States.

Dr. Kasinathan has professional engineering (PE) licenses from New York, North Carolina, and Florida and is a Certified Energy Manager (CEM), LEED AP Professional, Diplomat of American Academy of Environmental Engineers (DEE), and Registered Industrial Professional Hygienist (RPIH).

Dr. Kasinathan has been serving the territory pollution control industries for the past 35 years, assisting many universities and hospitals in New York, thus saving millions of dollars in bringing the facilities into complete compliance with USEPA and state and local agency requirements. He founded Environmental and Energy Solutions, Inc. and employs more than 20 people, currently assisting many hospitals in installing boilers, cogeneration facilities, and other sources.

He is a member of the American Institute of Chemical Engineers and the Air and Waste Management Association. Dr. Kasinathan has conducted several seminars about spill prevention, control, and countermeasure plans. He also conducts audits for air, waste, water, asbestos, and lead-based paint as per EPA regulations and provides corrective measures. He created compliance tracking software that is used by many facilities that interactively tracks their compliance with EPAEPA and city regulations, by which the facilities stay in continued compliance. His research on automated compliance efforts, utilizing sensors and controllers integrated with IOT, has been approved for U.S. patents.

Dr. Kasinathan has been a role model to many students aspiring to be facility engineers through internship and mentor programs. He has solved many work-related conflicts, adopted a diversity employment policy throughout his career, and improved productivity by increasing business growth. He has published several articles in environmental areas and has presented many seminars in business, productivity, and diversity.

Munikiran Mallu, MBA, is a project manager at Volvo Group Truck Technologies. Volvo Group's global truck technology and product development organization is responsible for technology research, engine development, product design, and all the technology and product development linked to truck operations, as well as supporting the products in the aftermarket.

A project manager with superior project management skills and a diverse technical background in trucking and off-highway industries, he is a Certified SAFe Agiist, Certified SAFe PO/PM, and Certified Professional Scrum Master I.

Munikiran earned an MBA at Fayetteville State University, North Carolina, and a BE in mechanical engineering at Madurai Kamaraj University, India, and he is a lifetime member of Beta Gamma Sigma (BGS) International Business Honor Society. BGS recognizes and honors the top 10% of undergraduate students, top 20% of graduate students, and all doctoral candidates from around the world in business schools accredited by the Association to Advance Collegiate Schools of Business (AACSB).

He earlier worked in different engineering roles supporting product development in Volvo, Caterpillar, CNH, and GE companies involving truck, agricultural, earthmoving equipment, and industrial appliances manufacturing industries.

He has worked with diverse people (races, ethnicity, ages, and gender) and understood how diversity helps bring ideas to the table.

Madalyn Bozinski is a senior environmental engineer at Environmental and Energy Solutions, Inc., a consulting engineering firm in Westchester, New York. She earned a BS in chemical engineering at Syracuse University's College of Engineering and Computer Science and a BA in political science at Syracuse University's Maxwell School of Citizenship in Civic Engagement, focused on environmental chemistry, public policy, and political organization and mobilization. She was also a member of the Renee Crown University Honors Program, in which she completed a thesis on mercury contamination in freshwater.

Prior to her current position, Madalyn was a process engineer at IBM's T. J. Watson Research Laboratory, specializing in photolithography. In her current work as an environmental engineer, she works with clients in the New York Metropolitan area to comply with environmental and other agency regulations from the NYC Department of Buildings (DOB), Fire Department (FDNY), Department of Environmental Protection (DEP), New York State Department of Environmental Conservation (DEC), and U.S. Environmental Protection Agency (EPA).

Madalyn is an alumna of Alpha Omega Epsilon Sorority, an inclusive network of diverse women with a passion for science, technology, engineering, and mathematics (STEM). She has also been involved in the Society of Women Engineers, the American Institute of Chemical Engineers, and the Tau Beta Pi Engineering Honor Society.

During her time at Syracuse University, Madalyn actively volunteered at a local elementary school with a mission of increasing culturally relevant learning for the diverse population of students in the city's school system. As a member of Alpha Omega Epsilon Sorority, she also spearheaded and participated in initiatives to encourage women to enter STEM fields and empower women in STEM.

Introduction

President Abraham Lincoln preserved the union by tapping the power of diversity. Rather than surrounding himself with a group of "yes" men who agreed with his every decision, he had advisers who could shed light on the complex issues facing the nation. His managerial functions allowed people the freedom to question higher authorities. In short, Lincoln's cabinet consisted of more adversaries than admirers because he recognized the possibilities of "diversity." On the contrary, unchallenged leaders of an organization have not had as much success, according to Survey of Leaders.

Diversified leaders allow ideas to evolve over time, if necessary. They are not focused on one way of thinking and will adapt their ideas to solve the issues at hand. The most successful leaders will use the power of diversity as leverage by listening to different perspectives, from diverse sources. They like to evolve their opinions as new information is gathered, but also know when to execute a decision. A true diversified leader will not be biased or hold any preconceived notions. They will not let prejudice or outdated ideas affect their decision making. Instead, they · will listen to the opinions and experiences of people of all genders, ages, religions, cultures, etc. A better solution can be found when there are diverse perspectives being discussed.

Abraham Lincoln was a pioneer of diverse leadership in the 19th century, but what does diversity look like in the modern corporate world? *Merriam Webster Dictionary* describes "diversity" as the condition of having or being composed of differing elements such as inclusion of different races or cultures in a group or organization. This includes programs intended to promote diversity in settings such as schools, organizations, and corporate businesses. Understanding that each individual is unique and recognizing their differences along the lines of race, ethnicity, gender,

DOI: 10.1201/9781003340966-1

sexual orientation, socioeconomic status, age, physical ability, religion, politics, or other ideologies is an integral part of diverse leadership.

From a global perspective, diversity enables more inclusive practices worldwide by understanding the range of differences that composes a group of two or more people in a cross-cultural and multi-national context. The modern corporate world must rethink diversity as the employee, customer, supplier, and partner configurations go global. Diversity can be divided into four categories from a management perspective: occupation, skills/abilities, personality traits, and value/attitude. Each industry deals with diversity issues differently depending on their respective corporate culture and associated management protocols. These industries are beginning to understand that diversity and inclusion are important parts of a sustainable future. As a result, many business leaders are beginning to create initiatives that foster a more diverse workforce. Companies that realize the importance of diversity and inclusion are in a better position to compete in the global market, by implementing policies and culture to support it. Furthermore, these companies have more flexibility because they have a wide range of talent to draw from. This results in greater employee retention, work relations, innovative work environment, and diversified supply chain.

THE NEED FOR WORKPLACE DIVERSITY

As the makeup of the U.S. population changes, companies are seeing the need to alter business and hiring practices from the way things have been run for decades or even centuries. A diverse workforce fosters innovation and creativity. Having workers from a wide array of backgrounds allows for vastly different experiences and knowledge to meld together, allowing for innovation and new ideas that never would have been fostered in a non-diverse group. Having working teams with diverse makeups and an accepting culture also empowers members of groups who have traditionally not been heard to feel more comfortable expressing their ideas and sharing their knowledge. Research by consulting firm Deloitte has shown that companies that require assimilation to company culture, or "cultural fit," rather than fostering diversity, are extremely detrimental to workers. Employees of companies that require "cultural fit" were found to have hidden aspects of their identity to fit in over 75% of the time. This number was closer to 95% for racial minorities. As a result, over 60% of respondents stated that this type of company culture was detrimental to their sense of self. Companies that corner themselves into one school of thought are effectively depriving themselves of growth and innovation, and hurting their employees in the process.

Aside from fostering a wide range of ideas, diversity also brings in a larger pool of skills. By building teams with individuals of varying backgrounds, there is a greater opportunity that different skills will be brought into the group. Creating teams with diverse skill sets not only equips them with a larger set of tools for creativity, innovation, and productivity, but it also encourages an environment where the entire team can form new skills. More experienced members of a team can contribute insights based on experience they have gained over years of working in an industry, while younger team members with less experience can offer an outsider's perspective that is not influenced by past experience. Team members with diverse backgrounds and different life experiences can also offer fresh perspectives and see situations differently than how others may approach a situation. In a team with different backgrounds and abilities, all team members can expect to gain or grow skills such as communication and problem solving.

Hiring a wide range of employees sets a stronger foundation for a business. Essentially, by fostering a more diverse and accepting workplace, candidates of all backgrounds will be more inclined to accept a position at a welcoming company. The more candidates that are interested in a company, the stronger the foundation of a business will be, because a wide range of candidates will bring a wide range of skills that a company can use to build a strong team that will allow for maximum success.

Research has shown that having diverse teams also leads to happier employees, which fosters higher rates of productivity. In a Deloitte survey conducted among over 1,500 employees in three companies with strong diversity and inclusion cultures, employees were found to be more productive, have lower rates of absenteeism, and have higher work/life balance satisfaction. Employees interviewed for this study were found to think of their company as successful. Studies have also shown that happier employees are 20% more productive. In short, if diverse workplaces increase employee happiness, then that happiness can manifest in increased productivity (Turner, Fischhoff, 2022).

Diversity especially manifests productivity if a company is diverse among both upper and lower management. Researchers from the University of Massachusetts Amherst, Vanderbilt University, and Florida Atlantic University found that companies that increased their diversity in both upper- and lower-level management saw productivity increase by up to $1,590 per employee per year for every one percentage (1%) of diversity increase (Cole, 2020). There are several factors that contributed to this result. This study showed that diversity is truly a "knowledge-based asset." It was found that diversity caused others to consider differing perspectives, leading to the development of more complex and thought-out approaches to the problems. These assets allow upper-level management to formulate stronger problem-solving strategies and lower-level employees to implement strategies with more tools and better skills. Having strong diversity in two interconnected groups of upper- and lower-level management also creates a more open and accepting environment, which encourages members of both groups to collaborate and share ideas more freely.

Conversely, a lack of diversity at upper and lower management levels was found to be detrimental to productivity. Companies that lack diversity at these levels were found to be 1.32 times less productive than those that had strong diversity. Another interesting finding of this study is that the most detrimental combination for productivity is having a diverse lower management without that diversity reflected in upper management. Companies with high diversity at upper-level management but low diversity at the lower level outperformed the inverse group by approximately double. This could be due to the fact that if upper management does not reflect the diverse makeup of their lower-level employees, these employees may feel less trusted and less willing to communicate with their superiors. The data found in this study is represented in Figure 1.1.

Diversity across a company is also an excellent tool for companies to create stronger relationships with their customers. By hiring employees from all backgrounds, companies can more effectively reach customer groups from those backgrounds as well. This allows room for the business to grow both economically and culturally. A company can only benefit from opening its audience to more people, and feedback from different groups of customers can help companies to evolve and innovate.

By encouraging diversity and inclusion, businesses are opening themselves up to a wider pool of talents that can only benefit their company. While a number of laws are in place to ensure companies are not purposefully using exclusionary hiring practices, some companies may still be losing out on the best talent by not having a strong enough focus on diversity and inclusion. Taking actions like emphasizing a culture of inclusion in job postings can attract a more diverse group of employees with a wider range of talents.

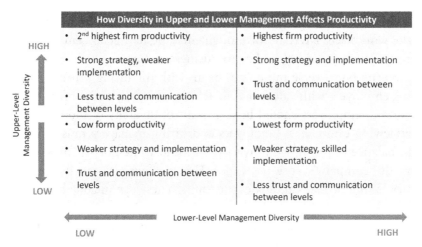

FIGURE 1.1   How diversity in upper and lower management affects productivity.
Graph compiled using data from Richard et al. (2021).

Aside from creating a more positive and accepting workplace for employees, implementation of serious diversity and inclusion programs has been proven to lead to higher revenues for companies. A study conducted by McKinsey Consulting found that diverse companies had the ability to achieve 19% higher revenues than their less diverse counterparts. Another study conducted by Bersin by Deloitte found that companies that have strong diversity programs saw a 2.3 times revenue increase per employee ("Diversity Wins: How Inclusion Matters", 2020). A major contributing factor to these revenues was found to be due to the greater innovation that diversity allowed. Companies that implemented strong diversity and inclusion initiatives provided programs such as education and training for underrepresented employees that fostered new ideas and stronger skill sets.

DIVERSITY OF THOUGHT

All of this combined allows for diversity of thought, which, now more than ever, is an essential part of business success. Even if a company has an extremely diverse group made up of different ethnicities, genders, and backgrounds, the company may not be fully maximizing its potential for diversity. Diversity of thought benefits both the group itself and the company and creates a more innovative dialogue. Individuals with different methods of approaching problems and varying perspectives can find more creative and optimized solutions to all kinds of problems faced in the workplace. Diversity of thought may be harder for employers to identify, but it could hold a key to business success.

Scouting out diverse minds can begin with the hiring process. While hiring employees with values that match the company's core principles is important, employers should be cognizant of diversity of thought when interviewing candidates. Employers and hiring managers must find the right balance of hiring with diversity in mind while not wholly straying from the company's core principles. Hiring managers can use various tactics to find diverse candidates for teams. These can include behavioral assessments and analyzing how a candidate reacts to certain scenarios. By performing exercises like this, hiring managers can assess whether a candidate would be likely to challenge an existing team or present information in ways that current employees may not. This would contribute to a team's diversity of thought (Training Industry Inc, 2022).

Diversity of thought can also be fostered among an existing team or workplace. One way this can be done is by the creation of interdisciplinary teams. When a project needs creative solutions, teams can be formed using employees with unlikely skill sets working together. For example, even though a construction project may be managed by contractors, architects, or civil engineers, bringing in other team members with knowledge of aspects like ecology, site remediation, regulatory processes, chemical treatment, or public health may bring perspectives and knowledge to a project that will allow a team to be more prepared for unexpected issues that could arise over the course of the work. Diversity of thought creates a breadth of knowledge that could be invaluable for a company or team.

Another strategy for fostering diversity of thought in the workplace is to use thinking tools. This could include giving personality tests to team members at the start of a project and pairing dissimilar thinkers to brainstorm solutions to a problem. It could also include brainstorming exercises like having the team list multiple correct answers to a question instead of agreeing upon just one, or having the team consider a problem from multiple different perspectives. One strategy that fosters diverse thought while also creating a welcoming environment, as discussed earlier in the chapter, is encouraging all team members to speak up. By giving everyone in the room a voice and an opportunity to be listened to, people will not feel like they need to fight to be heard; rather, they will feel that their opinion is valued. The other members of the team will also be introduced to a wider range of ideas.

While it is important for workplaces to foster diversity of thought, the best way to ensure this is by creating teams that are diverse in every aspect, not just in the way they think. Many have even argued that diversity of thought is merely a by-product of having highly diverse groups in terms of race, gender, ethnicity, or class. By cultivating environments with high diversity and inclusion, companies are opening a door to those with different backgrounds, perspectives, and ways of thinking.

WORKPLACE DIVERSITY IN THE UNITED STATES

Understanding and implementing diversity into the workforce will only become more important in the future. In the United States, the population is growing rapidly and becoming more diverse. World Population Review estimates the population of the United States to be 336 million as of 2023, and it is expected to grow to 373 million in the next 30 years. Not only is the population changing, but the racial profile is beginning to change as well. The U.S. population is currently about 60% White. However, as the racial and ethnic makeup of the United States becomes more diverse, the country is expected to have a White population of 48% over the next 35 years (Pew Research Center, 2018) .

Interestingly, as the U.S. population continues to grow and become more diverse with an increasing immigrant population, the unemployment rate is dropping. According to the Bureau of Labor Statistics, the U.S. unemployment rate was 3.6% in 2022, which is the lowest it has been in decades. The 2015 Census Report is predicting an increase in U.S. foreign-born population, from just 5% in 1965 to 19% in 2060 (World Population Review). Based on these predictions, leaders will have to embrace diversity if they want to have a successful future. There will be many challenges, as diversity-based discrimination is still a problem in this country, but this is an issue that corporations must face head-on to ensure their success.

The labor force in the United States is ever-fluctuating, which is another reason why companies should be welcoming to the idea of diversity and inclusion. The share of the U.S. workforce held by non-White Americans has risen significantly from 1979, from 11.7% to 22.3%, nearly doubling. The share of Hispanic and Latino Americans in the workforce has risen from 5% to 18%. Gender diversity has also increased, with women accounting for 47% of the workforce, up from 42% in 1979. Currently, 6.6% of America's workforce is over the age of 65, which increased from 3% in 1979. Conversely, the number of teens ages 16–19 declined from 8.2% in 1979 to 3.2% now. These shifts are likely due to the higher focus on college education and challenges many older Americans face with retirement.

Diversity will bring prosperity if it is embraced and leveraged properly in a corporate business setting. Slightly rephrasing what John F. Kennedy said in 1963, "My fellow workforce, ask not what your corporation can do for you; ask what you can do for your corporation." This is true for the corporation having social responsibility to the people of every social fabric of color, creed, race, gender, and language. Simply put, diversity in corporate business does nothing but bring additional prosperity. Therefore, it is imperative to add this diversity into the workforce. Diversity must be present in all forms – race, culture, religion, gender, age, ability, and socioeconomic status – to allow for true diversity of thought and, ultimately, business success.

Corporate businesses will face many challenges in the future, as the workforce becomes older and skilled labor becomes harder to find. To make things worse, past economic recessions have reduced the number of diverse subcontractors and vendors. It will become more important for businesses to pursue, attract, and train new, younger, and more diverse members of the workforce. If a company wants to sustain their success into the future, they must adopt diversity and inclusion. This can be achieved through creating initiatives that promote a more diverse workforce. A diverse workforce will better reflect and accommodate diverse populations within the United States. Diversity, as suggested by the Gardenswarz & Rowe four-layer model of diversity, is evolving beyond its initial form of racial diversity to include many other aspects of what makes individuals who they are. Besides race, it must be recognized that there is a large array of characteristics that make people who they are.

Race is no longer the only factor that corporate businesses need to take into account when approaching diversity. Diversity has grown beyond the traditional measures of race and culture. It must be thought about differently in the 21st century to include all aspects of human nature. Leaders of businesses and organizations must be continuously innovating as times change if they want to stay relevant and successful in the modern corporate world. They should foster a workplace culture that is young, innovative, inclusive, and diverse. This kind of inclusivity requires employers to be mindful of all the differences that may be present among their employees.

GENERATIONAL DIVERSITY

A new generation of young people has created an entirely new diverse group in terms of both skill sets and needs. While they are likely to be more familiar and skilled with technology, they may need a workplace that allows them freedom to use various types of technologies to complete their work effectively. New diversity attributes such as age, background, and social class can affect what employees value in their employers. These attributes can affect unlikely skill sets such as reasoning bias, the consumption and communication of information, and communication skills, and ability in areas like these tend to vary among different groups based on their life experiences.

Millennials, those born between 1981 and 1996, will make up approximately 75% of the global workforce by the year 2025. Within the next decade, this means that they will occupy most of the leadership roles in the workforce. With this new generation comes new perspectives on diversity. Millennials are also the most racially and ethnically diverse generation in the workforce, as shown in Figure 1.2. Older generations may understand diversity only as racial and cultural, but younger

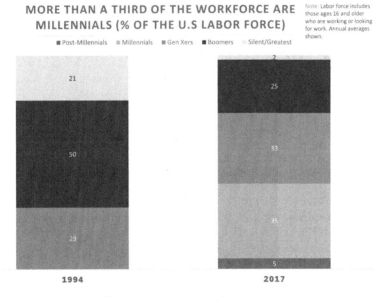

MORE THAN A THIRD OF THE WORKFORCE ARE MILLENNIALS (% OF THE U.S LABOR FORCE)

Note: Labor force includes those ages 16 and older who are working or looking for work. Annual averages shown.

■ Post-Millennials ■ Millennials ■ Gen Xers ■ Boomers ⬚ Silent/Greatest

FIGURE 1.2   Impact of millennials in the workforce.

Graph compiled using data from Pew Research Center.

workers may have a broader view to include not only identifiers such as gender and sexual orientation, but also through diversity of experience, background, and perspective. A core value of millennials is that the workplace is a supportive environment that allows for a wide range of perspectives and creates a feeling of welcoming for its employees, no matter their background.

A survey of millennials conducted by Deloitte in 2018 found that 74% of millennials believe a workplace culture of inclusion fosters innovation. Additionally, a survey from 2016 found that almost half of millennials are actively seeking out employers with good diversity and inclusion practices. Figure 1.3 demonstrates these findings. Ultimately, if companies want to attract the best potential workers coming out of this new generation of millennials, they must ensure their workplace allows for an inclusive, supportive, and welcoming environment. Not only will this bring in a larger pool of potential employees, but it will also create an environment that fosters innovations once they are working for the company.

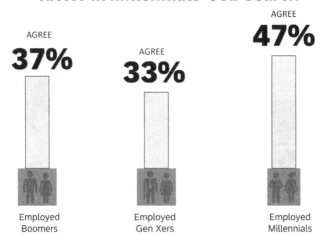

Diversity and Inclusion is a significant factor in Millennials' Job Search

AGREE
37%
AGREE
33%
AGREE
47%

Employed Boomers

Employed Gen Xers

Employed Millennials

"If I were to look for a new job tomorrow, a diverse and inclusive workplace would be important in my job search"

FIGURE 1.3   Importance of diversity and inclusion to millennials in their job searches.

Graph compiled using data from Weber Shandwick and KRC Research (Millennials at work: Perspectives on diversity & inclusion, 2018).

GENDER DIVERSITY

While progress has been made in the workplace over the past several decades, there is still work to be done toward achieving gender equality in many industries. Research has shown that women bring a significant positive contribution to companies. In fact, research has shown that companies with more women in their C-Suite are more profitable. Still, many companies have gender gaps, whether it is a wage gap, discrepancies in their hiring practices, or their general work environment. At a corporate level, women are severely underrepresented. Fewer women are hired than men for most entry-level positions, and from there, advancement continues to favor men.

All of these discrepancies exist even given the research of the significant contributions women make to the workplace. A Pew Research survey found that women are over 25% stronger in key business areas such as working out compromises, mentoring, honesty and ethics, and providing fair pay and benefits. From an economic standpoint, it has been estimated that closing the gender pay gap could increase the value of the global economy by $28 trillion. In a recent McKinsey report, companies with more diverse executive teams were found to be 21% more likely to experience above-average profitability. These teams were also found to be better suited to outperform their less diverse peers over the long term by 27%. Allowing for gender equity can bring a host of different skills that prove beneficial for companies, such as new perspectives on customer needs, improvements that can be made to products or services, and ideas on how the company can be improved in general.

While companies should feel a moral obligation to work toward gender equity among their employees, it is also just smart business sense. Companies should make strong plans to support and advance women. This would require a shift of norms in many company cultures, but companies can only stand to benefit from these changes. To truly give women equity in the corporate world, companies need to make serious changes like investing in training for employees and allowing for greater flexibility to fit work into employees' lives. By empowering women in the workplace, companies – and societies – will be more likely to grow and prosper.

Diversity factors such as age and gender are even more important to include when developing new markets. This diversity within the workforce will bring prosperity in the form of innovative new ideas. Research indicates that a diverse, inclusive, equal workforce will maximize human potential, so that all workers can advance and thrive. Creating this kind of culture is difficult and comes with challenges, especially on a global level.

WORKPLACE DIVERSITY WORLDWIDE

With advancements in technology and interconnectedness of globalization, teams with members from all over the world are becoming increasingly common. With Internet available almost everywhere, video networking, and the ability to chat with someone across the world in an instant, geographic barriers are no longer a hindrance to the workplace. This globalization has also created a host of multicultural and multinational teams that would not have existed otherwise.

As a result of globalization, new markets for opportunity have sprung up in what are known as the BRIC nations: Brazil, Russia, India, and China. Several leading U.S. corporations achieve more than 50% of their earnings from abroad, including IBM, Intel, and Dow Chemical. While multinational corporations have seen much success from global expansion, working with counterparts in several different countries can pose a challenge to U.S. multinationals. Most Americans have little exposure to the cultures of those in the BRIC nations. Teams must learn to work across different cultures, time zones, and languages. In particular, those in BRIC countries may have differing business values and cultural norms than Americans. For example, Americans use a very direct form of business communication, which is not common in most of the BRIC nations. Businesses in BRIC nations also tend to have a much more structured hierarchy than those in America.

Misunderstandings could certainly arise in multinational teams with these dynamics, which could make projects unsuccessful or teams ineffective. It is extremely important for a company to understand the nuances of assembling a multinational team. Cross-cultural training for management and workers alike can help ensure the success of these teams. With the proper training, teams with members from such diverse backgrounds could truly innovate in exceptional ways, but companies must be proactive about supplying these teams with the proper tools for success.

THE NEED FOR DIVERSITY IN INDUSTRY

While many companies have been making strides in implementing diversity initiatives and increasing diversity and inclusion for their employees, one field has been notably lacking. Industry, including construction, engineering, and most other STEM (science, technology, engineering, and mathematics) fields, have long been lagging behind other industries in terms of having diverse workforces. This book aims to specifically look into why diversity in industry is imperative, how it can be achieved, and the prosperity that can come from it.

Since its beginning, industry has been dominated predominantly by White males, and the majority of the industry workforce continues to be made up of that demographic. The percentage of women in engineering has risen only two points, from 12% to 14% since 1990. The percentage of women in computer science has actually decreased, from 32% to 25% since 1990. Those in various minority groups, including Black and Hispanic Americans, are also highly underrepresented in these fields. Nearly 70% of those in the STEM fields are White despite making up only 59% of the total U.S. population.

Lack of gender diversity is one of the major critiques of industry, specifically STEM fields such as engineering, as well as other fields like construction. Women are underrepresented in these fields, and they also face a significant gender wage gap. According to the U.S. Census Bureau, female professionals in STEM-related jobs make almost $16,000 less than their male counterparts on average. This gap also exists for Hispanic STEM professionals, with a gap of about $10,000 less than White STEM professionals. Women in these career settings are also 45% more likely to leave a STEM job as a result of a perceived hostile work environment.

A variety of factors have led to these statistics within industry. Many of these factors are systemic, and start in STEM education. Implicit and explicit biases have been shown to hinder minority students and women from entering careers in industry. One 2013 study showed that even with encouragement from parents and teachers, students in ethnic minority groups were persuaded against a career in STEM, citing discrimination as a major factor. Students who are discriminated against have tendencies to doubt their abilities, especially when it comes to STEM subjects like math and science. Conversely, studies have shown that diverse groups in academic settings have led to higher achievement and more positive outcomes.

As discussed at length in this chapter, the benefits of diversity can apply to industry as they would to any company. However, problem solving and innovation are some of the key tenets of fields like engineering and technology. These industries stand to benefit significantly from a diverse group of minds coming together to solve problems. These fields touch extremely important aspects of our lives, from food production, to chemical manufacturing, production of machinery, and even construction of our infrastructure. These key areas can only thrive when people work together, and as we have mentioned, diversity builds better, more effective teams. Breaking down barriers in industry to allow for more diversity is imperative, especially when considering a study by the National Center for Women and Information Technology. The study showed that mixed-gender teams had filed 40% more patents in the field of information and technology compared to all-male teams. People with different backgrounds and perspectives can approach problems differently in an effort to develop innovative solutions.

CONCLUSION

Now, more than ever, diversity in the workplace is a crucial aspect for ensuring business success. Not only is implementing diversity initiatives a moral imperative for companies, but much research has shown that diverse businesses will often become more successful. While many industries and companies have made strides in recent years to become more inclusive in terms of race, culture, and gender, there is still much work to be done. Additionally, the definition of diversity is ever-expanding. Hiring employees from all walks of life, with different backgrounds, values, and perspectives, has proven to lead to more innovation.

Companies must strive to achieve diversity at all levels. Having diversity at the employee-level but not the executive level, and vice versa, can actually be detrimental to a company, so it is imperative that diversity is achieved throughout a company. This widespread diversity also creates a better sense of welcome and inclusion for employees, a factor that has proven to lead to employee satisfaction and performance. The idea of inclusive company culture is also something that many employees are now actively seeking out in potential employers. Companies should be aware of these trends to ensure that they are attracting all possible candidates.

One area that specifically has work to do in terms of diversity and inclusion is industry – including engineering, construction, manufacturing, and other STEM fields. Industry has historically lagged behind in terms of diversity and inclusion. However, it is one of the industries that stands to benefit most from improved diversity and inclusion practices, as it is a field that relies on creativity and innovation. While this book will discuss overarching needs for diversity in business, a primary focus will be how these practices can improve and be implemented in industry.

Chapter 1 Review Questions

1. Describe the managerial skills of diversified leaders such as President Lincoln.

2. How should the modern corporate world rethink diversity?

3. What are the categories of diversity? How does each industry deal with diversity issues?

4. How does inclusion help to compete in the global market?

5. How does a diverse workforce foster innovation and creativity?

6. Does hiring a wide range of employees set a stronger foundation for a business? Why?

7. Discuss the relationship among productivity, employee happiness, and diversity.

8. State the reasons why corporations should consider inclusion at all management levels. Also, discuss how diversity in upper and lower management affects productivity.

9. What is the contributing factor to increase corporate revenue from a diversity perspective?

10. What are the percent shares of the U.S. workforce holding office positions?

11. Discuss the new perspectives on diversity from the millennials' (born between 1981 and 1996) point of view, including the impact of millennials in the workforce.

12. How will diversity bring prosperity in the industries?

13. How can you take advantage of an ever-fluctuating labor force in the global market to increase industrial productivity?

14. How can we achieve gender equality in any industry?

15. "Diversity of Thought," now more than ever, is an essential part of business success. Why?

16. It has been shown that technology tremendously helps achieve "Workplace Diversity Worldwide." Please discuss this in detail with statistics, including U.S. companies.

REFERENCES

Cole, B. M. (2020, September 15). *8 Reasons why diversity and inclusion are essential to business success.* Forbes. https://www.forbes.com/sites/biancamillercole/2020/09/15/8-reasons-why-diversity-and-inclusion-are-essential-to-business-success/?sh=26590bc01824

Diversity wins: How inclusion matters (2020, May 19). McKinsey & Company. https://www.mckinsey.com/featured-insights/diversity-and-inclusion/diversity-wins-how-inclusion-matters

Millennials at work: Perspectives on diversity & inclusion. Weber Shandwick. (2018, May 31). https://www.webershandwick.com/news/millennials-at-work-perspectives-on-diversity-inclusion/

Newkirk, P. (2019, October 10). *Diversity has become a booming business. So where are the results?* Time. https://time.com/5696943/diversity-business/

Pew Research Center. (2018, April 6). *More than a third of the workforce are millennials.* Pew Research Center. https://www.pewresearch.org/ft_18-04-02_genworkforcerevised_bars/

Richard, O. C., Triana, M. del, &, Li, M. (2021). The effects of racial diversity congruence between upper management and lower management on firm productivity. *Academy of Management Journal, 64*(5), 1355–1382. 10.5465/amj.2019.0468

Training Industry, Inc. (2022, March 22). *The importance of diversity of thought.* Training Industry. https://trainingindustry.com/magazine/may-jun-2019/the-importance-of-diversity-of-thought/

Turner, L., & Fischhoff, M. (2022, March 25). *How diversity increases productivity.* Network for Business Sustainability (NBS). https://nbs.net/how-diversity-increases-productivity/

Corporate Diversity Management

INTRODUCTION

The diversity makeup of a company is not something that is required by law but something much of corporate America practices on their own and thus is not regulated or standardized. Note that this is different from discrimination, which is highly regulated and will be touched upon later in the chapter. Corporate diversity management refers to applying diversity in all layers of management.

Typically, there are five layers of management.

- Entry Level

- Manager

- Director

- Vice President

- C-Suite

Corporate diversity is about broader depth, insights, and experiences to you as an individual and to your organization. No matter where you are on that spectrum (beginner to expert level), your organization's diversity management programs have the ability to say a lot about your company. The importance you place on diversity can enable your success. Diversity management programs tend to act as a metric for a company's success and ability to adapt in a changing corporate landscape.

Representation of diversity in all these layers is crucial in corporate diversity management. In the boardroom, there should be diverse

DOI: 10.1201/9781003340966-2

leadership for the prosperity of the company. Diversity is an inter-linking resource that can help a business grow and increase profits. In general, it goes beyond gender, race, and ability. When a company acknowledges the complexity of diversity and places value on these differences, organizations can thrive. Harnessing the offerings that a company's employees bring to the table will allow executives and the leaders within business to make greater accomplishments within any organization.

Ensuring diversity at a corporate level does not equate to simply en-acting affirmative action policies (AAP). AAP are procedures that can be implemented to consciously advance qualified minorities in the workplace. Managing diversity includes understanding a broad depth of insights and experiences that those at the topmost level of a company can utilize to bring people to their teams who sometimes may be over-looked. This can be done through actions such as placing values on unconventional skill sets, or broadening the idea of what proper quali-fications for a position could look like. This does not stop at the hiring process, however. Diversity management must be continuously im-plemented to keep an open mind and hear the input and concerns from all employees in an ongoing manner. This advice applies to all compa-nies, whether it be a large corporation or a new startup. No matter where a company lands on this spectrum, the importance a company places on diversity management programs can enable them to achieve success.

Harnessing the benefits of diversity must first be demonstrated by management at the top level of a company. Only then can a company truly tap into the benefits that diversity can provide. Diversity in a company is much more complex than gender, ethnicity, or ability. Management must understand that strong diversity in a company includes diversity of thought, experiences, and relationships. By placing values on all differ-ences that employees can bring to the table, previously unheard voices can provide new perspectives that can open the company up to unique ideas, which can help the company prosper.

With workplaces being such a large part of our lives, the social and cultural dynamics in the workplace tend to have the same complications as life outside of work. By having leadership that embraces the value of diversity in the workplace, a company can find renewed ways to foster important values among workers, including innovation, creativity, and even empathy for other ways of life. A company that does not make space for diverse views and where people are not exposed to the differences of others can stagnate a workplace and fail to help its employees grow both personally and professionally. However, embracing diversity in the workplace is not a simple task for leadership. Nurturing diversity in the workplace takes careful consideration from leaders at the top to ensure success and grow diversity into an asset.

Globalization has altered the dynamic of the workplace and the role of diversity in it. The idea of diversity and inclusivity not only encompasses gender, race, and ethnicity, but also religious differences, political beliefs, socioeconomic differences, sexual orientation, disabilities, and more. If handled correctly, companies can play a unique and transformative role in creating an inclusive society. This transformation can only begin from within an organization, and it needs a strong leader to foster this shift.

BUSINESS PERFORMANCE FROM DIVERSITY

Extensive research has shown that diversity can bring advantages to businesses such as

- higher profits,

- more creativity,

- better leadership,

- even stronger problem solving among workers.

This is due to the fact that the diverse backgrounds of employees can contribute a multitude of unique perspectives and experiences, helping to foster resiliency and creating a more effective work environment.

A 2017 study conducted by the Boston Consulting Group has even shown that companies that have taken efforts to make their management teams diverse have 19% higher revenues, resulting from the increased innovation that diversity can foster. In industries such as engineering and tech, where innovation is key, these findings are especially significant. These findings are shown in Figure 2.1.

Figure 2.1 depicts that there is a clear-cut difference (26% vs 45%) in innovation revenue between companies with below-average scores and above-average scores. Not only is it imperative for company leadership to promote diversity and inclusion internally, but they must also advocate for these same values to the public. One strong example of this is Proctor and

Companies with More Diverse Leadership Teams Report Higher Innovation Revenue

Companies with below-average diversity scores

Companies with above-average diversity scores

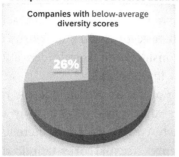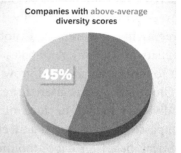

Average innovation revenue reported by companies

FIGURE 2.1   Benefits of diverse leadership teams.

Graph compiled using data from Consulting.us (2019).

Gamble's "We See Equal" campaign, which attempted to bring awareness to and fight gender bias by depicting boys and girls going against gender stereotypes. This reflects the company's internal work on this issue; 45% of Procter and Gamble managers and one-third of its executive board are women. Backing their public campaign with genuine action within the company is one of the reasons why Proctor and Gamble remains a strong and successful company.

Leaders in diversity can provide great examples of how to implement these policies, but diversity initiatives at every company will look different because diversity means something different at every company. Companies must tailor their diversity and inclusion initiatives to meet the needs of their specific organizations. It is important for companies to remember that diversity is not simply one initiative or one public campaign. Diversity and inclusion in the workplace should be a constant effort for a company to grow and improve, and this effort should be championed by its leaders. For genuine change to happen in a company, its leaders need to be fully invested in this kind of progress.

DIVERSITY LEGISLATION FOR EMPLOYERS

While being a diversity leader at any company can be of great benefit, it is also imperative that management understands the laws and regulations that are already in place that their company must adhere to when it comes to diversity and inclusion. The below list of legislation has been enacted in the United States over time that all employers must be aware of and must implement:

- **Title VII Civil Rights Act (1964)** – This act prohibits employment discrimination on the basis of race, color, religion, sex, or nation of origin. Employers cannot let these factors affect who they employ at their companies.

- **Pregnancy Discrimination Act** – The law prohibits against discrimination of pregnant people in any aspect of employment. Employers cannot factor an employee's pregnancy into decisions on hiring, firing, pay, work assignments, promotions, layoffs, and other benefits or terms of employment.

- **Americans with Disabilities Act (1990)** – This law prohibits discriminating against any individual with a disability in all aspects of public life, including in the workplace.

- **ADA Amendments Act** – This act ensures that disabilities under the ADA are defined broadly to provide maximum coverage for those benefiting from the law

- **Age Discrimination in Employment Act (1967)** – The law prohibits discrimination of employees over the age of 40, including disproportionately hiring young workers, providing less opportunity for advancement of older workers, and disproportionately laying off older workers.

- **Equal Pay Act (1963)** – This act requires that employers must pay men and women in the same workplace equal pay for equal work.

- **Employment and Reemployment Rights Act** – This act protects the employment rights of military service members and veterans, with provisions requiring that employers allow service members to return to their previous jobs and protecting against discrimination based on military or veteran status.

- **Civil Rights Act (1991)** – This act reinforces the 1964 Civil Rights Act.

- **Rehabilitation Act (1973)** – The law prohibits discrimination of disabled employees at federal jobs.

- **Genetic Information Nondiscrimination Act (2008)** – This act prohibits companies from using genetic information to discriminate against workers in all employment practices.

State and local governments may also have additional diversity legislation that employers have to comply with. This legislation at the federal and state levels makes discrimination unlawful in the workplace. Each of these laws discusses the responsibilities of both employers and employees to ensure discrimination is not present in various forms.

CHALLENGES OF DIVERSITY IN THE WORKPLACE

Diversity doesn't only have advantages; there are a few downsides, too. Those need to be addressed properly by the management.

Here are a few challenges:

- Too many opinions

- Hostility

- Communication issues

While we have discussed the many benefits of diversity in the workplace, there are challenges that company leaders and employees may face. Differences in language and culture can lead to misunderstandings or miscommunications among team members, which can lead to low morale (Carter, 2022). Implementing diversity can also be difficult, as some employees may be resistant to accepting those different from them. This can sometimes inhibit progress. It is also a logistical challenge for company leaders, as they must customize diversity to meet the needs of their specific organization. While there are challenges associated with implementing diversity in the workplace, the benefits and necessities of its implementation far outweigh the potential challenges.

DIVERSITY IN LEADERSHIP BY THE NUMBERS

A majority of companies have begun to strive for diversity in their leadership based on the values described above. However, it is important to understand that the idea of diversity bringing prosperity is not just an ideal; it has been proven through data. While many companies still have not met diversity targets in terms of management and executive positions, companies that have diversified their general workforce have seen impressive gains.

Why should this matter to corporate leaders? We found in Chapter 1 that if you are the only person like you in the room, you are more likely to leave your job and find something else. The environment and the

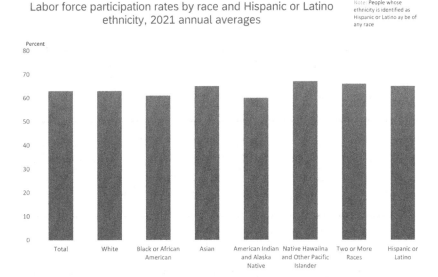

Labor force participation rates by race and Hispanic or Latino ethnicity, 2021 annual averages

Note: People whose ethnicity is identified as Hispanic or Latino ay be of any race

FIGURE 2.2   Makeup of U.S. workers by race and ethnicity.

Graph compiled using data from U.S. Bureau of Labor Statistics.

culture are less conducive to you feeling included. As managers, we spend a lot of money trying to recruit, sponsor, and retain people of diverse backgrounds, and yet we can get to a point where we are creating an environment that makes these people feel more excluded, not less. Figure 2.2 shows the disproportionality of race and ethnicity. Note the fact that diversity steadily decreases with more senior positions in a company.

The data that have been collected through surveys paint a picture of just how far things need to change before companies are truly representative of the makeup of society at large, and before salaries are comparable across categories like gender, race, ethnicity, and sexual orientation.

A study by McKinsey and Company showed that companies in the top 25% in terms of racial and ethnic diversity are over one-third more likely to see gains above the national industry median. Another study by Deloitte found that over a period of three years, companies with strong diversity saw over twice the cash flow per employee than their non-diverse peers.

IMPORTANCE OF MANAGING DIVERSITY

In Chapter 1, we discussed the importance of having diversity not just at the employee level of a company, but also at the top management levels.

Research has shown that diversity's benefits are most fruitful when the diversity of employees is also represented in the people who are managing them. Much of this has to do with feelings of alienation or prejudice if lower-level employees are diverse while their superiors do not reflect this diversity.

Without diversity at management levels, issues like stereotyping and prejudice can become destructive to a team or business as a whole. If managers cannot relate to the experiences or perspectives of their employees, the manager may filter their interactions with employees with a bias, even if their intention is not to be prejudiced. While forming and managing a diverse team may seem challenging at first, and may take longer to see results, the outcome will allow for innovation and higher performance from the entire team.

Studies by researcher Pamela Tudor at the Wharton Business School showed that managing diversity can be extremely effective once the team is dedicated to a common goal. By sharing an objective or task, diverse groups were able to overcome differences. However, this success is more difficult to achieve if diverse teams are not supported by management. Managers must be able to unite diverse employees over this shared goal. This unity cannot be achieved if the managers do not also embrace diversity in order to lead their team.

DIVERSITY IN INDUSTRY MANAGEMENT

As competition in today's economy increases, talent is crucial for increasing profits and improving a company's bottom line. Companies can only obtain the best talent pool by selecting from a large and diverse group of candidates. Several fields have yet to tap into the full potential of benefits that diversity can bring to organizations. A primary actor in this regard is industry – namely fields such as engineering, construction, and facilities management. While these fields are embracing diversity, it has been at a slower pace due to many factors unique to the industry.

The construction industry has increasingly begun to embrace the values that diversity and inclusion can bring. As recently as 2019, the Construction Management Association of America completed a long-term study and outreach program among the profession to set the course for a preferred future in the field of Construction Management. A priority found for this "preferred future" was to place high importance on diversity, equity, and inclusion. Having leaders in industry profess the need for diversity in the field is a major step and signal from leadership on the direction that the profession should be moving in. This type of active leadership is intended to set standards for management in industry to adopt as norms, allowing everyone in the industry to benefit.

The need for these actions toward diversity in the construction industry have also been highlighted by economic cases. Prior to 2020, the Bureau of Labor Statistics documented a trend of simultaneously low unemployment rates and low participation in the U.S. labor force (U.S. Bureau of Labor Statistics, 2023). The COVID-19 pandemic has only further highlighted a trend of global workforce shortage. An unprecedented shortage of talent is of particular interest to industries that rely on large workforces, such as construction.

The construction industry is expected to achieve approximately 21% growth between the years 2012 and 2022. However, a major factor affecting this growth is a workforce that is aging and an ever-smaller pool of skilled workers. This has driven construction managers to find creative ways to recruit new workers to the labor force. As the United States becomes more ethnically diverse – the 2020 Census found the percentage of the population made up of ethnic minorities to be steadily increasing – industries will need to attract workers from a much more diverse pool than in the past (U.S. Census Bureau, 2022).

This has presented a unique challenge in the construction industry because past economic recessions caused many independent contractors and vendors to go out of business, severely reducing diversity in the workforce. This challenge has not helped the construction industry, and has emphasized the need among industry leaders that embracing diversity and inclusion can only provide benefits to a workforce that has had many setbacks in the past few decades. Not only does embracing diversity help the industry recover from losses due to the 2008 recession and the COVID-19 pandemic, but it will also help the industry sustain growth as the country's population becomes larger and more diverse.

In recent years, the construction industry has begun to recognize that actively aiming toward a more diverse workforce will make the future of the industry more sustainable. Industry leaders have started initiatives to achieve a workforce with much more diversity and inclusion. In 2015, the Associated Builders and Contractors initiated an annual Diversity and Inclusion Summit to discuss pertinent issues in the industry. Many other industry leaders and professional associations have also followed suit with similar initiatives (Pfeiffer, 2015).

As seen in many other industries, management that not only embraces diversity in their fields, but also actively works to enact policies and instill culture to support it often see new success and growth as a result. Benefits of these initiatives include a larger pool of talent to choose from, higher employee satisfaction, improved work retention, and more innovative work environments. This type of success all starts when management takes the initiative to implement these policies and champion diversity and inclusion.

INDUSTRY INSIGHTS

Though diversity management in industry is improving, there are still many strides to be made. This is demonstrated in the following examples of insights from industry professionals at various levels of management.

According to Joe Smetona, an associate at the architectural/engineering/construction (A/E/C) advisory firm ECFG, diversity in the A/E/C industry is severely lacking (2023). Because of that companies are held back from achieving higher profits. Smetona addressed this fact as the keynote speaker at the National Conference of the Construction Management Associate of America. He asked attendees to take into consideration the slow pace of progress that the industry has seen in recent years. He punctuated this point by asking attendees to look around the room, implying a lack of diversity at the Association's own National Conference.

As regional manager of engineering at Gilbane Building Company, Linda Rosenberg has also provided her insight on the state of diversity in the construction industry, specifically the lack of women. At the executive level, the A/E/C industry has failed to include many women in high-level positions. Even women who do achieve executive success in this industry often have to wait longer or work harder than men who would be promoted to the same positions. She has noted that the industry is making strides in this regard, but there is still much progress. From 2019 to 2020, the number of women in the construction industry rose from 9.9% to 10.9%. This is progress, but the percentage of women in this field is still extremely disproportionate.

Both of these industry professionals found a common reason for the recent uptick in executive focus on diversity initiatives in A/E/C. The ongoing labor shortage in the industry opened executives' eyes to the value that a more diverse talent pool could bring. This revelation among industry leaders has motivated them to take real action, installing diversity and inclusion policies and initiatives at their companies.

CASE STUDY

EFCG, the A/E/C firm discussed above, surveyed 82 companies in the industry to gain further insights into diversity in the field. The companies surveyed encompassed 446,713 employees and a gross revenue of $120 billion (Byrne, 2022). The diversity figures that were uncovered were typical of the industry, but not ideal.

It was found that the average percentage of female employees at these companies reached about 30% between 2010 and 2015. The average percentage of employees who belonged to minority groups was around 15%. As of 2020, the percentage of female workers in the industry has stagnated at 27%, while the percentage of minority groups has risen to about 30%.

The study also found that, on average, firms with more employees tended to have a higher percentage of minority employees. AEC firms making gross revenue less than $100 million were found to have an average minority employee percentage of 8.9%. Firms with gross revenues over $1 billion employed minorities at an average of 22.6%. This trend among larger companies could be due to higher attention being paid to the issue of diversity. Companies with more resources and larger HR departments have more bandwidth to be able to spend time on developing and implementing diversity and inclusion policies and reaching out to a broader applicant pool when hiring.

However, even though there is some representation of women and minorities at A/E/C companies, very few of them hold high-level positions. When the EFCG survey was taken in 2015, none of the companies surveyed had a minority in an executive position. Another concern discovered in this survey was a gender wage gap, which increased with higher seniority levels. A separate study found that on average, women in construction management positions make 82 cents for every dollar a man makes.

"Companies' consideration of diversity & inclusion is not only important on the basis of values; it also has a material impact on their long-term performance," Barclays analysts said in a research report.

- Organizations in the top quartile with gender-diverse executive teams were 21% more likely to experience above-average profitability than their industry peers.

- Organizations with ethnic and culturally diverse leadership are 33% more likely to outperform their peers.

- A study by the Boston Consulting Group found that organizations with policies and practices that support diversity increased innovation revenue by up to 12.9%.

- According to McKinsey, 34% of C-suite jobs in the public sector are held by women and 22% by people of color compared to 21% and 15%, respectively, in the private sector.

CONCLUSION

As has been discussed in this chapter, diversity can breed creativity, increase innovation, and spell growth for any business. While some industries are making strides in this regard, others are still held back from achieving diversity and inclusion given a number of factors. No matter who you are or what your status is within your organization, we can all learn from one another. You can learn things that can benefit or even revolutionize your company. A major way to ensure a company implements diversity and inclusion to ensure success is to have strong leadership that is committed to this effort. Corporate management can make or break the diversity at a company by dictating priorities and choosing where to divert resources. If those in positions of power are not focused on diversity goals, then the initiatives will never gain traction at a large level. We will discuss how these efforts can be implemented in later chapters, but as a rule, diversity and inclusion start from the top at most successful companies.

Chapter 2 Review Questions

1. Discuss the layers of corporate management.

2. What is corporate diversity?

3. Do affirmative action policies (AAP) satisfy the corporate diversity requirements?

4. How do you manage diversity in industry?

5. How do you benefit from harnessing diversity?

6. What are the complexities when managing diversity in the workplace? Describe with examples.

7. What are the advantages brought by diversity in the industry?

8. Discuss the revenue growth in an industry that is fully compliant with diversity in the workforce.

9. What are the benefits of a diverse leadership team?

10. Discuss the diversity legislation, if any, for employers.

11. Are there government-mandated or suggested legislations for diversity in the workplace?

12. What are the challenges of diversity in the workplace? Discuss in detail.

13. Discuss industry insights for diversity management.

14. Discuss the survey results conducted by companies such as of ECFG and A/L/E/C on diversity insights.

15. Discuss women's role in corporate diversity management.

REFERENCES

Byrne, S. (2022, February 23). *Women in construction.* NEIT. https://www.neit.edu/blog/women-in-construction

Carter, C. (2022, June 8). *Top challenges of diversity in the workplace.* Fraser Dove International. https://www.fraserdove.com/challenges-of-diversity-in-the-workplace/

Consulting.us. (2019, June 11). *For US businesses, management diversity means financial, innovative success.* Consulting.us. https://www.consulting.us/news/2359/for-us-businesses-management-diversity-means-financial-innovative-success

Corporate diversity: If you don't measure it, it won't get done (2020, March 6). McKinsey & Company. https://www.mckinsey.com/capabilities/strategy-and-corporate-finance/our-insights/corporate-diversity-if-you-dont-measure-it-it-wont-get-done

Diversity, Equity, & Inclusion | Construction Management Association of America (2023, February 27). https://www.cmaanet.org/about-us/diversity-equity-inclusion

Peiffer, E. (2015, October 21). *Can focus on diversity keep construction industry from a sinking bottom line?* Construction Dive. https://www.constructiondive.com/news/can-focus-on-diversity-keep-construction-industry-from-a-sinking-bottom-lin/407657/

U.S. Census Bureau. (2022, June 10). *2020 Census illuminates racial and ethnic composition of the country.* Census.gov. https://www.census.gov/library/stories/2021/08/improved-race-ethnicity-measures-reveal-united-states-population-much-more-multiracial.html

U.S. Bureau of Labor Statistics. (2023, January 1). *Composition of the labor force.* U.S. Bureau of Labor Statistics. https://www.bls.gov/opub/reports/race-and-ethnicity/2021/home.htm#:~:text=Labor%20force%20participation,-Among%20the%20race&text=The%20participation%20rate%20for%20Asians,Created%20with%20Highcharts%2010.3.

Diversity Trends and Challenges

I n the previous chapters, we have discussed a pressing need for diversity of all kinds at all levels of a company. Countless research has shown the variety of benefits that diversity can bring to a business. Some businesses and industries have had more success achieving diversity than others. This chapter will examine how diversity can be implemented throughout a company by examining current trends in some of the most diverse businesses. We will also discuss challenges faced by companies that have struggled to achieve diversity in some or all forms. As discussed in Chapter 2, industry, including fields like engineering and construction, has faced some of the largest challenges in many areas of diversity.

CURRENT TRENDS

Leaders in diversity have learned to adapt as the definition of diversity has grown and evolved. Ultimately, those with success in diversity have learned that the definition of what a diverse company looks like is ever-changing. One of the key markers of a diverse company is the willingness to adapt and innovate to meet the needs of its employees and consequently enhance the productivity of its employees. In this section, we will look at specific considerations that companies can make to allow for accommodations that help diversity flourish within a company.

Remote and Hybrid Workforce

Due to the COVID-19 pandemic, remote work became a mainstay of the United States and global workforce. While various industries require in-person work, a large swath of the workforce is fully remote or working only a few days outside of the home. Approximately 58% of the U.S.

DOI: 10.1201/9781003340966-3

workforce has the option to feasibly work from home at least one day per week, and 87% of employees take opportunities provided by employers for these flexible work schedules.

Remote work and hybrid schedules give new growth opportunities to individuals who may not have been able to take advantage of them in a non-remote position. People who have to shoulder most of the responsibilities for household and family care duties, most often women, have vastly more opportunities in the age of remote work. By eliminating commute time, these individuals are given more time to complete their household responsibilities and also have the flexibility to watch over children who may be attending remote school while still being able to satisfactorily complete their job duties. Remote work can also allow these family members to be closer in case of emergencies, rather than potentially far away from their children or families at an office building in another town. Several million women actually had to leave the workforce in 2020 if their job did not provide a remote option, as they now had increased responsibilities at home. Hybrid and remote work has allowed these individuals, specifically women, to be more successful in both their work and home lives, and understandably, studies have shown that women have a stronger desire to work from home than men.

Hybrid and remote work options have also been extremely beneficial to many with physical or mental disabilities. For those with physical disabilities, a long commute to the office could be a barrier to being able to get to a job, and may rule out an opportunity altogether. For those with mental disabilities, being able to work from home can mean some workers are more productive in a comfortable setting where there are fewer distractions or people. Individuals could have any number of disabilities that do not affect their job performance but that a commute or in-person work environment can negatively impact.

People with economic and housing limitations have also seen benefits from remote work options. In many cases, people cannot afford to live in the city or town where they work. This can bar people from many job opportunities if they live too far from major hubs or it can decrease employee happiness by forcing those who do not have the means to live near their work to have to commute several hours every day. By not allowing for remote work, location becomes a barrier to entry and can then rule out a workforce that is socioeconomically diverse.

However, when remote work is involved, a new set of diversity accommodations must be considered. As mentioned above, oftentimes women are expected to balance home and work responsibilities disproportionately. While working from home can be extremely beneficial in this regard, employers must also be sensitive to the fact that this configuration may call for the need for caregivers to work outside of typical business hours to accommodate for other responsibilities. Those with disabilities may also need certain accommodations when working remotely. While remote work can have challenges, it has allowed for expanded opportunities for many disadvantaged groups. Employers should take this into consideration when thinking about how they can expand diversity and worker satisfaction at their companies.

Allowing for Gender Identity and Expression

Recently, awareness of different gender identities and expressions has been at the forefront of society. One of the main goals of diversity in the workplace is to create a space for employees to work their best and achieve a company's goals. By allowing for inclusive best practices like accommodating gender-neutral restrooms and using inclusive language to make all employees feel welcome in the workplace, the company is only setting itself up for more positive outcomes and a better employee experience.

Multigenerational Workforce

As briefly discussed in previous chapters, having a multigenerational team can bring some very unique strengths and talents to a project. A Randstad study found that 83% of teams felt they were able to come up with more innovative solutions to problems when they were part of an age-diverse team. For example, younger people have in-depth experience with modern technology and digitalization, while older employees can bring strong expertise and experience. When specialties such as these are combined, quality ideas backed by expertise can be done extremely efficiently with the use of technology.

Having employees span over many generations, it allows for the easy transfer of knowledge, skills, and even job positions. When an employee is ready to retire, having a younger colleague who has been brought up under the retiring employee allows for a seamless transition and opens a new door for the younger employee. This type of generational pipeline can help a company save on hiring costs and allow for efficient transitions.

However, there can be some challenges with a multigenerational team. At the outset, it may be very difficult for older generations to communicate effectively with their younger colleagues, and vice versa. This is under-standable, being that the preferred communication styles of different generations can vary greatly. Younger generations prefer to communicate over messaging and emails, while older generations often prefer phone calls or face-to-face contact.

Employers must learn how to facilitate relationships between workers of different ages to reap the benefits of these multigenerational teams. It is important that leaders use different strategies to accommodate and celebrate their employees of all ages and generations. The best way to facilitate solid intergenerational working relationships is to keep an open mind. By not making preconceived judgments and being open to having constructive conversations with those who may have different view-points, a relationship can be built based on mutual respect.

Hiring Diversity Professionals

To help workplaces navigate many of the diversity trends and new prac-tices, it is often advantageous to hire a diversity professional. These in-dividuals are committed to implementing the ideals of diversity and inclusion in the workplace. They combine skills like human relations, hiring, and leadership to ensure that companies can meet their diversity goals in all aspects of the business. Diversity professionals will typically hold educational sessions for employees on various diversity and inclusion topics. They are trained with skills to talk to employees about things like creating a culture of inclusion and using inclusive language, introducing and implementing new diversity initiatives, and even dealing with human resources complaints regarding issues such as discrimination or harass-ment. These individuals have training to implement and encourage diversity that a typical management or executive may not have and can bring this skill set to teach a company how to meet their diversity goals.

Supporting Mental Health

Another major impact of the COVID-19 pandemic is the effect it had on people's mental health. In turn, this has significantly impacted the workforce. The closing of businesses, loss of jobs, adjustment to working from home, lack of socialization, and constant health concerns are just a few of the factors that severely impacted mental health for workers across the country and across the world.

Mental health is an aspect of diversity and inclusion that can often be overlooked. However, everyone's mental health needs are different, and employers need to take this into consideration to ensure the mental well-being of all their employees. Companies will need to intentionally work to support mental health for all employees and understand that some may need more support than others. With recent issues in the country surrounding minorities and women, employees in disadvantaged or underrepresented groups may face a significant mental burden. Employers should understand the additional stress of these factors and be mindful of the effects this could have on their employees. Not only should employers implement policies around mental health, but they should also be mindful of how supervisors and coworkers behave and empathize with one another.

SHORTFALLS OF DIVERSITY

While there are a wide variety of benefits from implementing diversity and inclusion in the workplace, allowing for a diverse and inclusive space can come with its own set of challenges. Because of some of these challenges, companies and even whole industries can often fall short in terms of diversity.

Communication Barriers

One of the benefits of having a diverse workforce is that it contributes skills and perspectives from all different cultures around the world. However, many of the individuals that can contribute these things may not have grown up or be accustomed to colloquialisms in the country where their company is. This can lead to regional or national differences in language that can cause misunderstandings. Certain gestures or sayings in one culture can be perfectly benign while causing offense to someone with a different understanding of its meaning. The differences in perspective that diversity brings, while great for innovation, also have the possibility of causing tension among interpersonal relationships. This does not mean that all diverse companies are bound to have these issues. In fact, as long as a company fosters an environment of cultural sensitivity, these issues can be avoided or easily resolved.

Stereotypes and Prejudice

In today's society, unfortunate as it is, stereotypes about certain groups or types of people still exist, and some people still hold prejudices against those who are different from them. While an employer may try to ensure that there is no place for those who discriminate in their company, there is a chance that some employees may still hold prejudiced biases or believe in certain stereotypes. In a work setting, some employees may use these biases to justify being uncooperative or even hostile with another employee or group of employees. This could cause dysfunction among teams, which can disrupt the goals and benefits of forming a diverse team in the first place.

Lack of Trust

When diverse teams exist in a workplace, it is important for a manager to be sensitive to how members of minority groups may feel about a group dynamic. Oftentimes, unconscious biases can contribute to team members in minority groups to feel that they are being treated unfairly in comparison to their colleagues who are not in minority groups. Managers need to be especially sure not to give preferential treatment to any team member, especially those in a majority group. This could cause significant tension and lack of trust among minority team members. If minority group members notice this, they may feel uncomfortable to speak up and instead become quietly frustrated or unhappy. This, too, can affect the outcomes of the entire team's work. Managers must be understanding of these possibilities and make sure they have the proper leadership and communication skills to make every member of their team feel included and valued.

Visa Sponsorship and Accommodation

Building a diverse team with employees from all walks of life means that there is an increased chance that some employees may not be citizens of the country in which they work. Oftentimes, an employee working on a visa will want to apply for citizenship to be able to continue working in a country that they have built their career foundations in. In order to ensure that international employees can stay with a company past their visa expiration, companies need to dedicate time and resources to helping their employees gain citizenship or extended visas. A company should have a standard policy and guidelines when it comes to dealing with visa arrangements. They must also make sure these policies comply with labor laws as well as immigration laws. If a company plans to sponsor the citizenship of an employee, they must be prepared to expend time completing administrative forms and paperwork and have a sizable budget for the visa or citizenship application process, which often includes lawyers, application fees, and travel (Olson, 2022).

DIVERSITY SHORTCOMINGS IN STEM AND INDUSTRY

As touched upon in Chapter 2, industry including STEM fields and construction have faced many shortcomings in terms of diversity. Nearly half of all science and engineering jobs are held by White males. With the dominance of a field to that degree, those in minority groups can often be overwhelmed and uncomfortable, leading to less enthusiasm for those in minority groups to want to pursue careers in these fields. This disparity can start extremely early. One study of young African American girls found that this group was actually more interested in fields like science or math than their White counterparts, but only 66% felt welcome in the scientific field compared to 80% among their White counterparts.

Much of this disparity has to do with the culture surrounding male-dominated fields like science, engineering, facilities, or construction. Because the field is so heavily dominated by males, there is a persistent masculine culture in these fields that can make women feel unwelcome or out of place. This can lead to effects like imposter syndrome, in which some people feel like they are not capable or do not deserve their place in an engineering field, even though their performance shows otherwise. If this lack of confidence among minority groups and women is cultivated early by a male-centered culture, many women or minorities may be extremely discouraged from even pursuing a career in a STEM or industry field ("STEM has a Diversity Problem").

Another shortcoming in industry is how those in minority groups have reported being treated in these fields. Many have reported facing discrimination in the workplace, noticing a workplace culture that is insensitive to their needs, and a feeling of favoritism to those in majority groups. A Pew research study conducted in 2018 highlighted many of these issues. Black individuals working in STEM fields were four times as likely as their White counterparts to report that their company did not place a significant enough focus on increasing racial and ethnic diversity. They have also reported facing a significant amount of racial discrimination in the workplace, specifically compared to their counterparts working in non-STEM jobs (a 12% increase in the STEM field). As shown in Figure 3.1, 62% of Black Americans, 44% of Asian Americans, and 42% of Hispanic Americans reported experiencing discrimination in the STEM field.

This issue runs even deeper in other areas of industry such as construction. As of 2020, 88% of construction workers were White, and only

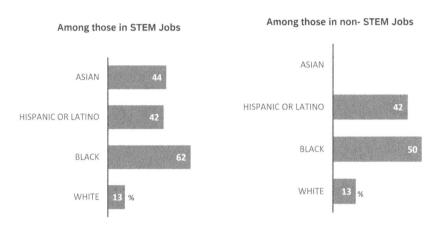

FIGURE 3.1   Individuals in STEM reporting discrimination.

Graph compiled using data from Pew Research Center (Anderson, 2018).

10% were women (McKissack, 2018). Similarly, the trend of minority groups experiencing discrimination is also increasing in the construction industry. A survey by the National Institute of Building Sciences shows that nearly all minority groups have faced a significant amount of discrimination or prejudice while working in construction. While all minority groups faced these challenges, Black or African American respondents faced a disproportionate amount of discrimination, with 72% reporting these experiences. Gender discrimination was also heavily reported by 66% of respondents. Instances such as these can help explain why these industries have faced such difficulties in recruiting or retaining diverse workers (Funk, Parker, 2019).

It is important to look at why these industries in particular face this current state of diversity. Much of this has to do with systemic problems that have been built into the structure of these industries for decades. Some issues start at the top with leadership – those recruiting new employees, growing business, and shaping companies' cultures. For a workplace to be inclusive, leadership must actively work to make it that way. Leaders can shape an inclusive culture through both social and budgetary means. If leadership is not committed to changing the social dynamics of a company to be more diverse and inclusive, they likely will not allocate company resources to these efforts either.

Due to the current trends in diversity mentioned at the start of this chapter, many leaders in industry have created diversity and inclusion policies for their companies, but few have truly gotten into the idea of creating an inclusive workplace. The same study that showed over 70% of Black workers faced discrimination in the field of construction also found that 43% of the respondents' employers had some sort of diversity and inclusion initiative on record. Based on these statistics, many in industry are not making strong enough efforts to implement the policies they have written.

Often, it may initially be uncomfortable for both leadership and employees to change company practices from the way things have always been done to push a diversity and inclusion initiative, and that is why many companies will avoid taking real actions to implement these policies. Making a good faith effort to implement diversity inclusion can involve developing meaningful and well-enforced anti-harassment and discrimination policies. Workplaces with these policies should have zero tolerance for employees who do not comply to set a standard for the company that no forms of harassment or discrimination are allowed. Companies should also look at restructuring policies such as how promotions are awarded, paid time off, and parental leave. Including training programs regarding important diversity and inclusion topics such as unconscious bias, cultural sensitivity, and allyship, as well and creating a worker safety culture that is also focused on issues like mental health are other ways that companies can make true good faith efforts at effective diversity programs. If industries like construction and engineering want to gain the benefits of a diverse workforce, they must think beyond technical skills and industry knowledge and truly embrace diversity culture.

Another downfall that could be contributing to a lack of diversity in industry is not practicing diverse recruiting and hiring. Especially in technical fields like engineering, companies tend to hire a similar type of candidate, and often engineering and technology companies tend to recruit even from the same colleges. Recruiters should work to expand who they recruit and make meaningful partnerships with diverse institutions such as Historically Black Colleges and Universities (HBCUs) and technical schools, which may have more students from underserved communities. This issue is similar in the construction industry as well.

One study conducted on the construction of Terminal 5 at the London Heathrow Airport found that many difficulties were faced finding workers for this project because those running the project only used their traditional recruiting practices, meaning they did not look beyond their existing workforce and reach out to more diverse communities that could have supplied more workers.

Creating diverse hiring practices can take many forms. In industries like construction, companies can work with groups such as unions or apprenticeship programs to attract new workers. They can also network within their communities by sponsoring events or awards and scholarships. Diverse hiring practices also include actions such as using inclusive language in job descriptions, reviewing resumes without names attached, or "blind," to remove unconscious bias in the review process, and creating and advertising an inclusive company culture starting with the onboarding process.

One of the biggest challenges still facing industry, as touched upon previously, is ongoing challenges of workplace harassment. Not only have these issues kept minority workers from wanting to join the industry, but they also create an extreme health and safety risk.

SUCCESSES OF DIVERSITY IN INDUSTRY

While challenges may exist specifically in STEM and construction fields, several businesses in this industry have found success through implementation of diversity initiatives. One of these success stories is manufacturer Johnson and Johnson, who ranked first on Diversity Inc. Top 50 list in 2018. Out of ten members on its board of directors, two were black men and three were White women. There are also four women, including two women of color, on the company's executive team. Johnson and Johnson also held community outreach programs for mentoring of diverse students by industry leaders.

By restructuring their employee benefits to include same-sex domestic partners, Johnson and Johnson also became the 17th most LGBTQ-friendly company. Uber is the number one LGBTQ-friendly company. It has created an extensive program to promote LGBTQ equality in various cities across the country.

Google is one company that has a truly multinational and global workforce. Because of its massive reach, it has had to adapt and increase accessibility for those all over the globe. Google's embrace of a global economy has led to its success, with 80% of the global market share for search engines. The company also made it a priority to reach as many users as possible. They realized that to reach the largest market possible, they must speak the language of their customers. While there are over 4,000 languages spoken in the world, Google found that they can communicate with 99.3% of users by translating into 40 languages. However, they did not simply translate their search engine. Google created the 40-Language Initiative, which looked to understand cultural differences in the languages they were translating to make their site as accessible to all users as possible.

Daimler-Chrysler found success implementing diverse teams in product development. They actively work to make heterogenous product development teams and have found the best results from these teams. The company works to ensure that a team has an equal number of male and female members, a varying age range, and that no more than half of the group is of one ethnicity.

These companies have all achieved success in diversity in different ways and using different strategies. A study conducted by the European Union found three areas in business where diversity seemed to matter most. Each case listed above incorporated diversity in one or more of these areas:

1. Marketing – A company has diverse projects that cater to their consumers. When diverse consumers see that a company reflects their needs, trust is built between the consumer and the business.

2. Operations – Having a diverse workforce and truly valuing diversity in the workplace can actually save a company money. By creating an inclusive culture where all employees feel welcome, rates of turnovers and absenteeism will be reduced. Conversely, inclusive culture will also encourage employees to be engaged in their work.

3. Innovation – We have discussed at length how diversity can be an asset to a team. Having team members who represent various target markets can make a company more in-tune with their customers and allow for new ideas that a homogenous team may not be able to offer.

REMEDYING DIVERSITY SHORTCOMINGS

It may seem difficult for a company to overcome its diversity shortcomings, especially industries that are entrenched in old systems and norms. However, as evidenced above, this shift is possible and extremely beneficial. To begin this shift, companies should change a few key mentalities, beginning with hiring and going all the way through day-to-day work practices (Yvanovich, 2021).

Shifting from Culture Fit to Culture Add

When hiring, many companies look to recruit employees who will fit well into the existing company culture. However, when the company culture is flawed and a business is looking to alter its practices to allow for more diversity, those in hiring positions should not look for exact "culture fits." Instead, they should focus on hiring employees who will add positive attributes to the existing company culture, but not necessarily fit into a preconceived idea of what an ideal candidate for the role would be. By shifting focus from a typical "culture fit" candidate, the company becomes more welcoming to those with new ideas and different perspectives, adding to innovation. Changing this mindset in the hiring process allows for the hiring of candidates from a multitude of different backgrounds, which will then create the groundwork for fostering diversity throughout the company.

Taking a Stance on Diversity Issues

Companies should also demonstrate to all their employees that they are aware of cultural or societal issues that may impact their lives. It is no longer a best practice to remain neutral on issues of diversity. In recent years, when turmoil over gender, racial, and identity inequality has consistently been a topic of discussion and contention in the news and in society, companies sitting on the sidelines can cause more harm than good. Employees in minority groups who are affected most by these issues will feel that the company does not understand their struggles or is indifferent to them. This can be detrimental to fostering an environment of inclusion in the workplace, making workers feel unappreciated and affecting team morale. Those leading companies should be aware of the cultural issues that could be affecting their employees, and they should make a point to take a stand on issues and give a message to their employees that they will be supported in light of what may be going on in the world.

Making Use of Diversity

Once a company has achieved a diverse workforce, they need to continually work to maintain it by utilizing their employees in the best way possible. In diverse organizations, employees will have a broad set of strengths and skills that may be at odds with one another. For a company to make the most of their diverse teams, management must know how to structure teams that will highlight everyone's strengths.

EMPLOYEE-INITIATED DIVERSITY

While much of the responsibility for diversity programs rightly falls on the employer, there are several initiatives that can be taken by employees, both diverse and not, to effect change internally. Sometimes, diverse employees will need to be their own advocate, especially if a company is not taking strong-enough steps. Diverse employees can seek out mentorship from those in the company who have similar backgrounds, and more accomplished employees can even make an effort to voluntarily take on mentorship roles. One major reason minority groups are often shut out of high-profile careers is due to lack of access to education. While not possible in all cases, advocating for oneself and even approaching an employer about further educational opportunities can allow diverse employees to get a leg up in their careers. Most importantly, a person's actions are a major factor to what they can achieve in their career. By advocating for oneself and striving for success through actionable items, all individuals can excel. Still, many underrepresented groups face barriers to success in most workplaces, and companies need to understand how to address these issues to ensure success for all of their employees.

CONCLUSION

As with any new endeavor, implementing diversity initiatives in the workplace can present challenges. The workplaces and employees' standards of what is expected from their employers are ever-growing and evolving. Companies must be well-equipped for the challenges they may face and understand the strategy to overcome them. While implementing effective diversity strategies may seem like a struggle at first, the ultimate positive outcomes for both the company and its employees will be extremely beneficial.

Chapter 3 Review Questions

1. What are the current trends in diversity management?

2. During the COVID-19 pandemic, how was diversity accommodation considered and implemented?

3. How is diversity connected with gender identity and expression?

4. Explain the hiring process of diversity professionals.

5. Mental health is a concern for diversity and inclusion. Should the industries take this aspect into consideration, and if so, what can be done to mitigate this crisis?

6. Describe the shortfalls of diversity and how to mitigate the shortfall, including those in the STEM industry.

7. Discrimination, especially for Black STEM workers, is experienced in both STEM and non-STEM industries. Explain how many leaders created policies to solve this problem.

8. If a company fails to take action to implement non-discriminatory policies, what are the steps to be taken for compliance?

9. Do recruiters play a vital role in hiring a diverse workforce? Discuss the case of the London Heathrow Airport Construction project.

10. Discuss workplace harassment and its impact on the company's predictivity.

11. Will inclusion of same-sex and the LGBTQ personnel in the workforce help benefit the companies? Provide examples.

12. Google's success is truly attributed to multinational and the global workforce with multiple spoken languages. Language plays a vital role in diversity management. Elaborate this concept with the examples of other companies.

REFERENCES

Anderson, M. (2018, January 10). *Black Stem employees perceive a range of race-related slights and inequities at work*. Pew Research Center. https://www.pewresearch.org/short-reads/2018/01/10/black-stem-employees-perceive-a-range-of-race-related-slights-and-inequities-at-work/

Funk, C., & Parker, K. (2019, December 31). *4. Blacks in STEM jobs are especially concerned about diversity and discrimination in the workplace*. Pew Research Center's Social & Demographic Trends Project. https://www.pewresearch.org/social-trends/2018/01/09/blacks-in-stem-jobs-are-especially-concerned-about-diversity-and-discrimination-in-the-workplace/

McKissack, D. (2020, October 29). *Diversity: What's needed to move forward.* https://www.enr.com/articles/50522-diversity-whats-needed-to-move-forward

Olson, N. (2022, March 17). *4 Things standing in the way of diversity in construction.* Safesite. https://safesitehq.com/diversity-in-construction/

STEM Has a Diversity Problem (n.d.). ParentMap. https://www.parentmap.com/article/stem-stream-diversity-education-statistics

Yvanovich, R. (2021, July 28). *5 Challenges of diversity in the workplace and how to overcome them.* https://blog.trginternational.com/challenges-diversity-workplace-how-to-overcome

Diversity Management and Implementation

D iversity in all levels of management is a must for better outcomes and performance of the company. There are typically three layers of management in industry. They are as follows:

1. Lower Management

2. Middle management

3. Senior Management

DIVERSITY IN LOWER-LEVEL MANAGEMENT

Lower-level management comprises managers who oversee the groups that work on the core engineering work. Different background managers will bring diversity in the company where engineers will interact every day. This layer of management is quite visible for all the employees, so the company should focus on including strong diversification in lower-level management.

Group managers play a pivotal role in securing the right workforce who executes the core work of new projects or operations. There should be norms in this layer to hire from different backgrounds, ethnicities, and demographics. This is very critical when hiring people to make a globalized product or local product that suits the taste of a global market.

DIVERSITY IN MIDDLE-LEVEL MANAGEMENT

Middle-level management represents the chief engineers who manage the total departments such as an electrical department or chassis department, in a motor vehicle company. In addition, there will be

 DOI: 10.1201/9781003340966-4

operations, aftermarket, and manufacturing divisions in this layer. This is the main link between senior management to lower layers of management. There should be good diversification in this layer. This way senior management can see diversified people in their meetings. Otherwise, senior management cannot be so aware of all diversification in lower layers of management.

Chief engineers control the overall departments. They are the ones who delegate powers to the group managers in running their respective groups. In general, any employee can be visible to two layers of management for the most part. A strong mix of diversification in this layer of management can have a positive impact on employees who see their backgrounds being understood by those above them.

DIVERSITY IN SENIOR MANAGEMENT

This is the topmost layer of management in the management pyramid. They control the overall company's decisions and policies. This is the combination of C-Suite titles and board of directors who control Chief Executive Officers. Diversification in this layer proves to other layers of management that the company is truly dedicated to diversification and inclusion. Ensuring diversity in senior management sets an example for all at the company that importance is placed on a diversified team, and gives a model for other teams at the company to strive for.

The senior management team is also the most public-facing. Therefore, it is necessary to have diversification in this layer. Having a board of directors from all diversity classifications will give a competitive edge to the company. Decisions in this layer will influence the future, profitability, and direction of the company.

Therefore, diversification in all three layers will have a positive effect on the company outlook. Better products can be designed due to healthy debate over the designs and operations. Brainstorming sessions will be much more meaningful for competitive output. In general, senior management should not preach about diversification but should show that in the selection of the board of directors.

ACHIEVING DIVERSITY IN LEADERSHIP

How can a company achieve diversity at different levels of management? The answers vary from each company and each management level. It is clear that for diversity to have its desired benefits, a company must be diverse at all levels. A study conducted by McKinsey showed that there is a statistically significant relationship between diversity at leadership levels and improved financial performance. Companies that have executive racial and ethnic diversity have been found to be one-third more likely to have profits above national averages. This has also shown true with gender diversity, which correlated with a 21% increase in average profits.

Despite these benefits, diversity at all levels of management is still a struggle for many companies. While some progress is being made, and companies are increasingly adding women and people of color to their executive boards, there is still much room for improvement. A 2020 study conducted by Mercer showed that 85% of executive positions are held by White individuals. Figure 4.1 shows that diversity steadily decreases as career level increases, leaving little diversity at the highest levels of management.

It is imperative to look into what can be causing these major diversity gaps in leadership and how to fix them. There are many causes cited by diverse professionals for the struggle of achieving diversity in leadership.

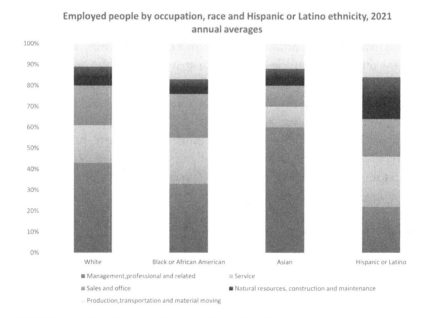

FIGURE 4.1   Distribution of race and ethnicity by career type.

Graph compiled using data from U.S. Bureau of Labor Statistics https://www. bls.gov/cps/cpsaat11.htm.

Some have found that minority employees lack facetime with senior leaders and miss out on opportunities to build relationships with those responsible for promotions. Systemic lack of access to college degrees can also be hindering minority groups from rising quickly in companies. Many in marginalized groups also believe that unconscious bias of employers have taken them out of the running for promotions or high-level positions by being deemed not to be a "cultural fit" or unprepared for high-level roles.

While this trend may be worrisome, there is much that a company can do to support leadership development for minority workers. One strategy is to implement bias training and hold management accountable for their hiring and promotion practices. We have previously discussed shifting away from the concept of hiring a culture fit, and rather hiring those who will add to the company's culture, even if they do not perfectly blend with the existing culture. Making these changes can be extremely beneficial because even if management is not knowingly acting on assumptions or bias, research has shown that people are naturally more comfortable with people who come from similar backgrounds and look and act like them. Since a majority of management is still White and male, this unconsciously causes disadvantage to women and people of color.

Another way to address these biases is by taking the subjectivity out of activities like hiring and promotions. Companies should implement clear standards for promotions and apply them consistently. Everyone in the company should also be aware of these standards so they can hold management accountable for implementing the standards. When hiring, employers should ensure that hiring teams are made up of diverse leaders. If a diverse candidate is not suitable for a position based on experience or skill, the hiring team should clearly communicate these deficiencies to the candidate so the individual can understand the areas that they can improve upon to be eligible for a promotion or position in the future. For internal candidates, a company can even make an effort to create a plan for that employee to move upward by providing actionable goals for the candidate to achieve.

Once diverse candidates are hired, the work toward inclusion is just beginning. A company must have a strategy for creating a sense of inclusion for diverse team members. One way to do this is by creating various programs to incorporate new diverse team members and set them up for success. Mentorship programs and leadership development training for minority team members can help to foster a sense of belonging and a feeling that the company truly wants these employees to succeed. Over time, as diversity increases in leadership, these programs can become larger and led by diverse leaders who can relate to the experiences of the employees they are training and mentoring (Walker, 2020).

MANAGEMENT IMPLEMENTATION OF DIVERSITY

Achieving a diverse makeup across a company is one small portion of a larger diversity strategy. Once a company is diverse at all levels, management plays a key role in actually implementing diversity and inclusion initiatives. Companies hoping to have true diversity and inclusion should make it a priority to use some key methods of implementation from the executive level to the employee level.

Sense of Belonging

It is important that executives and managers are aware of how employees feel about the workplace. One key factor of fostering diversity at all levels of management is to make employees feel welcome and to make sure they feel that they belong at the company. Creating a sense of belonging for employees is not a simple task. This goes hand in hand with the company having a true culture of diversity. A company that is diligently working to create a diverse and inclusive culture will also be working toward achieving a sense of belonging for their employees. The ultimate goal is to create an organization of individuals that makes employees feel accepted and like they can be themselves.

Leadership with Empathy

Oftentimes, diversity and inclusion implementation is left to the hands of a company's HR department. However, diversity policy will not go very far if the entire company, and especially leadership, does not participate fully. Executives, senior, middle, and lower management must all take steps to buy into the diversity policy of the company for it to be successful. This is not an easy task, and leadership must be equipped with knowledge of how to actively create a culture of inclusion.

One of the main strategies that management can use to further the company's diversity initiative is leading with empathy. Management needs to be able to connect with the needs and values of the company's diversity initiative. They must truly understand and appreciate the importance of the initiative in order for it to be implemented. By turning to empathy, they can put themselves into the shoes of employees that the diversity initiative is designed to benefit. Managers must think back to times when they were excluded or undervalued in the workplace, and understand that this is how many of their diverse employees may feel. By being in touch with the need for diversity initiatives, empathy can create a stronger feeling of necessity for these initiatives even among those who may not directly benefit from them.

A Holistic Approach

While diversity implementation at the management level is a necessity, it should not be the only way that diversity is implemented. A top-down approach for diversity that is being enforced by management may create a feeling of burden rather than commitment among employees. Every employee, not just leadership, must understand their role in the company's culture to be able to understand how different employees experience their workplace. By understanding the nuances of the company's culture, leaders should then identify which ways to implement different diversity initiatives. Some strategies may work better with a bottom-up approach, allowing lower-level employees to give their insights, while others may work best if middle management oversees certain initiatives. This type of leadership knowledge is learned through experience, but having leaders who are cognizant of the actions that need to be taken is an important first step.

Moving Past Quotas

While a diversity initiative may first begin with an effort to hire or promote those from minority groups, this is not the only way to achieve diversity. Often, when companies implement quotas such as these, they see that they have hired more minority individuals and believe that the rest will work itself out. This is not the case, and achieving true diversity and inclusion takes much more effort. To ensure that diverse hires remain at the company, employers must make a daily effort to create working conditions that promote inclusion. They must also devise metrics for success of diversity and inclusion programs to ensure that the measures they have taken truly do improve the company's culture.

Promoting inclusion must be considered in every decision made by a company's leadership. Simple aspects like who gets invited to meetings and how often individuals are able to share input have a large impact on the feeling of inclusion in a workplace. Company leadership should work to create an environment where everyone feels welcome and where their unique contribution feels valuable.

Ongoing Practice

Training employees on inclusivity strategies is an important step toward an inclusive workplace. However, once employees learn techniques for fostering a welcoming environment, they must be able to put these skills into practice in order to form new habits. These new habits can only be formed if the environment within the workplace supports open conversation and positive tension that allows employees to practice skills of inclusion. All members of a company can foster this kind of environment. One way to do this is to appoint leaders among non-management employees within the company who are responsible for inclusion implementation in their own groups. Having a dedicated inclusion leader who is not part of management can create an environment that encourages change that is not seen as a directive from a superior but rather as a constructive conversation with a colleague.

Emphasize Positive Connections

One issue that employers should be wary of when implementing diversity and inclusion initiatives is the presence of fear. When people's beliefs are challenged, this can lead to fear and distrust among employees. The challenges of diversity in the workplace should be framed positively to avoid instilling fear and causing further division. Instead of only focusing on areas that need to be improved, companies should also highlight and celebrate achievements in diversity and inclusion that their teams have made.

BRINGING MANAGEMENT ON BOARD

Management implementation of diversity initiatives is critical to the mission of these programs, but it is not always easy to get these actors on board. It is especially imperative for the participation of middle management, who have the largest influence in terms of employee interaction, recruiting, hiring, and promoting employees. Without enthusiasm from mid-level management, diversity initiatives often cannot get off the ground, even if there is strong enthusiasm from senior leadership (Johnson, 2023).

Because mid-level management oversees most daily interactions among employees, these individuals have one of the most crucial roles to play, as they have many opportunities to champion the company's diversity and inclusion initiatives through these interactions. They have the power to foster a sense of belonging and welcoming for employees and to create an environment where people feel safe to offer opinions and perspectives.

Research has shown that managers who work to strengthen diversity and inclusion initiatives create an environment where minority employees feel their work experience is free from bias. Figure 4.2 shows that minority employees with mid-level managers who consistently support diversity initiatives are one and a half times as likely to feel that their daily work experience is free of bias, and are twice as likely to agree that they do not see obstacles to diversity and inclusion in recruiting, hiring, and advancement at their workplace. These results are not comparing companies with no commitment to diversity and inclusion. Rather, these drastic results are seen between companies that only have a strong executive commitment to these initiatives versus companies with a strong executive commitment plus frontline managers who are consistently committed to diversity and inclusion.

The difference frontline managers can make is tangible. For instance, when a manager is a strong supporter of unconscious bias training, they can learn and implement strategies directly in their team. For instance, in the past, administrative tasks for group meetings such as ordering lunch or taking notes primarily fell on women due to unconscious bias. With a manager who is aware of these dynamics, a new system can be implemented such as rotating administrative tasks among all team

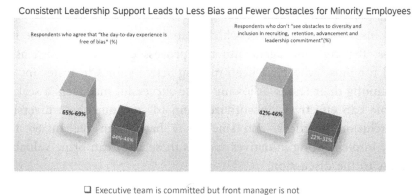

Consistent Leadership Support Leads to Less Bias and Fewer Obstacles for Minority Employees

Respondents who agree that "the day-to-day experience is free of bias" (%)

65%-69%

44%-48%

Respondents who don't "see obstacles to diversity and inclusion in recruiting, retention, advancement and leadership commitment"(%)

42%-46%

22%-31%

❑ Executive team is committed but front manager is not
❑ Consistent leadership commitment

FIGURE 4.2   Effect of consistent leadership commitment by frontline managers.

Graph compiled using data from Taplett (2021).

members. Not only does a strategy like this remove bias, but it also distributes the work fairly and evenly, which may not have been the case previously.

There are a few key strategies that executives can use to get middle management on board to ensure their diversity and inclusion initiatives are successful:

Include Managers When Creating Diversity and Inclusion Initiatives

When companies are creating diversity and inclusion initiatives, it is important that all stakeholders are a part of this process. While bringing in external consultants may be helpful to identify the company's areas for improvement, it is important to keep those managers who will actually be responsible for implementing these initiatives as part of the process. By including managers in this process, they can also feel a sense of ownership and pride over the strategy that they have created, and may be more likely to enthusiastically implement it.

Including managers in this process can also be extremely beneficial because they can provide working knowledge of daily operations and interpersonal dynamics that executives do not often get to see. These managers also understand which aspects of the company can be strong areas for improvement in terms of diversity and inclusion. They know the day-to-day diversity issues being faced and can help create initiatives and programs based on real-life scenarios rather than hypotheticals. This kind of input will allow for a smooth transition and implementation of the new initiatives because the real-life applications will be incorporated into the plan itself.

It is important to consider which managers to include in these initiatives. Bringing managers into this process can also be seen as a reward for those who have been making active efforts to foster inclusivity among their teams. Allowing these successful managers a seat at the table can effectively incentivize them to continue these diversity and inclusion practices, and since they have been recognized for already implementing initiatives such as these, they can bring valuable insights to the discussion.

Communicate Expectations

Much of the reason why some diversity and inclusion initiatives fail may be because managers are unaware of what is expected of them. If managers do not understand their role in these new policies, it can be extremely hard for any part of the initiatives to be implemented. By clearly laying out how managers are expected to enforce new inclusion policies and utilize diversity strategies, they can feel more comfortable that they are correctly implementing the company's initiatives. If management feels confident in what they are doing, they are more likely to continually execute the company's diversity and inclusion goals. One way to communicate these expectations is for executives to clearly define goals to be achieved. This will give management a target to meet and an incentive to achieve an inclusive environment.

Accountability

Another way to ensure management implements diversity and inclusion initiatives is to hold them accountable. If every effort is made to include management in the process of creating diversity initiatives and their expectations for implementing the initiatives are clearly communicated, then it is fair to hold managers accountable for ensuring these policies are being put to use among their teams consistently. For example, if a manager is expected to hire a new candidate, upper management should hold the manager accountable for ensuring their hiring process is free of bias and makes an effort to be as inclusive as possible. This can be done by working with the manager to ensure they write an inclusive job description or helping them to create diverse hiring criteria.

Another way to hold management accountable for implementing these initiatives is to include diversity and inclusion in their performance reviews. If managers are aware that their efforts to incorporate diversity and inclusion practices will factor into their performance reviews, they are likely to put more of a focus on these efforts. Intel implemented this idea at their company by putting in place monetary incentives for those who helped the company achieve its diversity goals. The company doubled their referral bonus for employees if they referred candidates from minority groups. Because of this incentive, Intel met and even exceeded its goal for diversity in hiring.

Continuous Training and Ongoing Support

One mistake that many companies make with their diversity and inclusion plans is putting in work at the beginning, but not maintaining the effort to make a lasting impact. Companies must make a long-term commitment to their diversity and inclusion plans, and much of this responsibility lies in the hands of management. Giving management a budget and opportunities for ongoing diversity training can be the key to a long-lasting diversity and inclusion program.

It has been shown that one-time diversity training sessions do not leave the desired lasting impact on employees. Instead, management should be creative in keeping these initiatives alive. One unexpected strategy is to make training sessions voluntary. Research has shown that when employees make the decision to attend these sessions on their own, they are more receptive to the message being taught in the training.

Maintaining an ongoing form of training is also an important aspect of maintaining these initiatives long term. Instead of one-time training sessions, companies should consider weekly training sessions or offering mentoring programs. By making recurring training sessions, they can be more focused on single topics and managers can then have time between sessions to absorb and implement what they have learned.

STRATEGIES FOR MANAGING DIVERSITY

Once management is brought on board to implement diversity and inclusion initiatives, they must figure out how to manage a diverse team on a daily basis without forgetting the core messages of their diversity training. Managers who hire and mentor employees are some of the key players in boosting diversity in a company, and they must utilize this leverage to help their company succeed. However, many managers may feel overwhelmed by these new responsibilities or claim they do not have time to focus on implementing diversity initiatives. As we have discussed, failing to implement diversity and inclusion initiatives can actually end up costing the company money.

After a manager has received diversity training and understands their role in implementing the company's initiatives, they may feel overwhelmed and not know where to start. The following strategies can help outline a plan for management to oversee diversity initiatives and put them into practice (Reiners, 2021).

1. Compile Diversity Data – A manager should understand the makeup of their team. By collecting data on the diversity of the team, the manager can understand a diversity baseline and take into consideration areas that can be improved.

2. Inclusive Job Listings – Once a manager understands the diversity gaps in their team, whenever positions become available, they should

make an effort to fill roles by encouraging a wide range of applicants. By creating job listings that use inclusive language and are open-minded about the required skills that could be needed for the job, a larger number of applicants, and more from diverse backgrounds, may apply to the posting.

3. Mentorship for New Hires – Creating more inclusive job postings should ultimately lead to the hiring of more diverse team members. Since these new hires may come from non-traditional backgrounds compared to the company's existing employees, it is important to ensure retention of these workers by creating a welcoming and accepting environment. One way to help with this is through mentorship programs. By recruiting experienced employees who may also be from minority backgrounds to help these new hires adjust to the company culture, it creates a sense of community and belonging and increases the likelihood that employees will feel positively about the work environment and want to remain at the company.

4. Provide Leadership Training – One pattern that managers may fall into is failure to successfully promote diverse employees. Many times, this stems from these employees not being exposed to opportunities that would allow them to be viable candidates for promotions. Managers must be aware of this and make an effort to help their diverse employees get the experience they would need for a promotion. This would include proactively looking to assign them work that would help them gain experience or by enrolling them in programs such as leadership training.

5. Prioritize Diversity and Inclusion – It can be easy for managers to become overwhelmed with the idea of implementing a diversity and inclusion initiative on top of their other job tasks. However, it is important for them to prioritize these initiatives and make sure they do not fall to the back burner. The best way to ensure a diversity and inclusion program succeeds is by having a manager who is enthusiastic about its implementation and is dedicated to ensuring its success.

6. Set Goals and Metrics – Even if managers are enthusiastic about implementing diversity and inclusion initiatives, they may not be doing it efficiently or effectively. The best way to stay on course with implementing these strategies is to come up with a system of

metrics to measure diversity success. By collecting diversity data at the beginning and using this as a baseline to measure achievements, managers can see their progress through real data and clearly see areas that have succeeded and areas that can be improved upon.

CONCLUSION

Implementation of diversity and inclusion initiatives is a critical task for employees at all levels of management. Each level of management is responsible for different aspects of the success of an initiative. While executives and upper management are responsible for creating the diversity program and taking high-level actions for its conception, mid-level managers are the individuals who can have the largest impact on whether these programs succeed. Executives and upper management are responsible for bringing mid-level managers on board to support diversity and inclusion initiatives. However, mid-level management must make efforts to ensure that the programs are being focused on in both the short term and long term of the company.

Chapter 4 Review Questions

1. How will different levels (low, middle, and senior) of management help bring a diverse workforce?

2. Discuss the C-Suite's responsibility for diversity inclusion.

3. Describe the results of the 2020 Mercer study of Diverse Decline situations?

4. Describe the distribution of race and ethnicity by career level.

5. Discuss the reasons expressed by minority employees for the struggle to rise quickly in companies. Hint: education.

6. Will the equal opportunity promotional policy help continue the diversity process in the industry? Provide an example.

7. Explain, with examples, mentorship and leadership training for minority team members.

8. Discuss the sense of belonging in the workplace.

9. What is leadership with empathy in a diverse process?

10. Is a holistic approach necessary in diversity management?

11. Is the quota system in the hiring process considered under-employment? Discuss why.

12. Describe positive connections.

13. Describe the effect of consistent leadership commitment by frontier managers.

14. Describe the communication of expectations.

15. Discuss the strategies of managing diversity in industries.

REFERENCES

Johnson, I. (2023, February 1). *Want a successful DEIB strategy? Get middle management buy-in - ChartHop - All your people data in one place.* ChartHop. https://www.charthop.com/resources/blog/dei/get-middle-management-buy-in-for-dei/

Reiners, B. (2021, October 20). *Diversity management: A guide to strategies and best practices.* Built In. https://builtin.com/diversity-inclusion/diversity-management

Taplett, F. B., Garcia-Alonso, J., Krentz, M., & Poulsen, M. (2021, September 7). *It's frontline leaders who make or break progress on diversity.* BCG Global. https://www.bcg.com/publications/2020/frontline-leaders-make-break-progress-diversity

Walker, L. (2021, February 6). *Are we there yet? Bridging the diversity gap in executive leadership.* https://www.linkedin.com/pulse/we-yet-bridging-diversity-gap-executive-leadership-lisa-walker/?trk=public_profile_article_view

U.S. Bureau of Labor Statistics. (n.d.). *Employed persons by detailed occupation, sex, race, and Hispanic or Latino ethnicity.* U.S. Bureau of Labor Statistics. https://www.bls.gov/cps/cpsaat11.htm

Importance of Diversity in the C-Suite

A C-Crew is the group of executive-level managers within a corporation responsible for high-level functions of the company. Their titles start with letter C, for "Chief," as in Chief Executive Officer (CEO), Chief Financial Officer (CFO), Chief Operating Officer (COO), and Chief Information Officer (CIO). The C-Crew members work together to ensure a corporation stays true to its established plans and policies and is deemed the most important and influential group of individuals within a corporation. These C-Crew members have a plethora of experience and finely honed leadership skills and have changed the world with their contributions. Many C-Crew executives relied on functional know-how and technical skills to climb the lower rungs of the corporate ladder and have cultivated more visionary perspectives needed to make sound upper management decisions. Below are the roles and titles of C-Crew members:

- Chief Executive Officer (CEO) – The highest-level corporate executive serves as the face of the company. The CEO consults other C-Crew members for advice on major decisions. CEOs can come from any career background, if they have cultivated substantial leadership and decision-making skills along their career paths.

- Chief Financial Officer (CFO) – The CFO position represents the top of the corporate ladder for financial analysts and accountants striving for upward mobility. The CFO must have the ability to manage the company's financial portfolio, manage its accounting, and research investments for the company. CFOs have insights on

DOI: 10.1201/9781003340966-5

global markets and work closely with CEOs to source new business opportunities while weighing the financial risks and benefits of each potential venture.

- Chief Operating Officer (COO) – The COO ensures a corporation's operations run smoothly in areas such as recruitment, training, payroll, legal, and administrative services. The COO is usually second in command to the CEO.

- Chief Information Officer (CIO) – The CIO is a company's leader of information technology. Typically, a CIO will have experience as a business analyst and will use these skills to excel in the position while developing technical skills in disciplines such as programming, coding, project management, and mapping. CIOs are usually skilled at applying these functional skills to risk management, business strategy, and finance activities. In many companies, CIOs are referred to as the chief technology officers.

SENIOR MANAGEMENT DIVERSITY

Diversity increases the bottom line for companies, and increasing the diversity of leadership teams leads to more and better innovation and improved financial performance. Companies that have more diverse management teams have 19% higher revenue due to innovation. Initiatives envisioned and spearheaded by a company's C-Suite can lead to the creation of a positive and diverse working environment, but only if paired with strong policies such as equal pay and an inclusive company culture.

Building a diverse leadership includes practices for holding top leadership accountable, and strategies for changing organizational cultures. When looking at the most successful companies in terms of diversity, most have achieved success by having an engaged executive team who supports the mission of diversity. Senior executives are often a company's prime messengers, so their encouragement of diversity programs and the reinforcement of diversity messaging is imperative. Companies with significant diversity success have had direct involvement from their CEOs, including promotion of events and initiatives such as diversity reviews to ensure management is implementing the diversity strategy as part of the company's business model. There are several ways that a C-Suite can push diversity initiatives to ensure that the company's goals are achieved.

1. Finding and Developing Diverse Employees

   Many times, a company's diversity initiative fails to flourish because the company has no effective system in place for developing leadership among its diverse employees. Companies can bring in large numbers of diverse employees, but that cannot be the last step. They must invest in programs that give these employees the skill sets needed for strong leadership. Strong employee development programs, specifically for diverse talent, will increase the likelihood of employee retention and company success.

   The role of the C-Suite is to ensure a company's leadership development program is strong. While 77% of companies have official leadership development programs, more than half of these companies' CEOs and HR departments do not believe that the goals of these programs are being achieved (Barrington, Troske, 2001).

2. Employee Support and Mentorship

   Another important aspect that should be spearheaded by the C-Suite is the provision of mentorship programs and professional support for diverse employees. Having strong professional networks and the support of mentorship within a company, especially mentorship by diverse leadership for diverse employees can avoid non-inclusive cultures and feelings of alienation from diverse employees. While typically middle management would handle the implementation of these programs, it is the C-Suite who can make or break their success by providing funding and resources.

3. Training and Education about Gender/Race Equity

   Diversity training increases the awareness of the demographic profile of an organization and challenges any negative preconceptions employees may have regarding minority groups. Research by Tel Aviv University professor of sociology, Alexandra Kalev, has found that mandatory diversity training is the least effective method for increasing diversity in management and, in fact, can even be counterproductive. When employees are forced to take one-off diversity training without company support of a larger diversity culture, this can often be seen as a performative and ineffective exercise. Leaders in the C-Suite can mold that company culture to a holistic one valuing diversity in all aspects to achieve real buy-in to the initiatives from its employees.

4. Manager Accountability

Organizations that are most successful in achieving managerial diversity have human resources systems and practices that hold managers and executives accountable for achieving diversity objectives and encourage them to actively develop women and people of color. Measurement tools like 360-degree feedback to peer reviews, employee attitude surveys, performance reviews that incorporate diversity objectives, and periodic reviews of workforce demographics directly or indirectly link diversity to management bonuses and incentives.

5. Diversity in Senior Leadership

An organization with more diverse representation in senior management will likely achieve greater profits. Organizations in the top 25th percentile for gender diversity on their executive teams were 15% more likely to experience above-average profits. Most recent data show that this number has grown to 21%. Figure 5.1 shows how strongly this diversity is correlated to profitability.

As with much data of this nature, the profitability varies by region and company health, but based on these data, a significant link between profitability and both gender and ethnic diversity seems to exist. While gender diversity seemed to have the smallest impact, the evidence suggests that gender diversity has increasingly marked potential profitability over time. Ethnic diversity, at the employee, executive, and Board of Directors level also showed an increased chance for profitability.

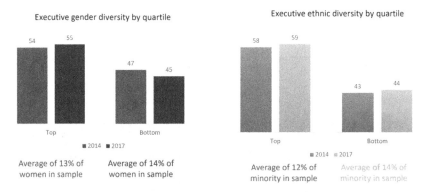

FIGURE 5.1   Gender and ethnic diversity correlated to profitability.

Graph compiled using data from McKinsey & Company, Diversity Wins.

MAKING A POINT TO CLOSE THE LEADERSHIP GAP

A perceived lack of diverse leaders at the executive table within firms of all sizes is a recurring theme. Cultural and ethnic diversity on the board level will have a big impact on profit goals. Among U.S.-based firms, Black Americans held only 4% of senior executive positions, even though they represent 10% of all university graduates in the country. Hispanics and Latinos – 8% of college graduates – also held 4% of all senior executive positions at U.S.-based. Asian Americans held 5% of those types of roles, while representing 7% of college grads. In the United Kingdom, Black and minority ethnic people make up 22% of all university graduates, yet among the U.K. companies in the study only 8% of all executive roles were held by members of that group. In the United States, women of color represented the smallest portion of executive roles, with Black, Hispanic, and Asian women making up roughly 30% of all female-held executive jobs. White women made up the other 70%. By country, Australia had the largest share of women in executive roles, at 21% among all the firms, and 30% representation among board members. In the United Kingdom, women held 15% of executive jobs and 22% of board seats. Figure 5.2 shows the representation of women of color on executive teams in the United States.

Mentoring programs have been found to be a major benefit when it comes to developing a diverse executive team and Board of Directors to make up a company's C-Suite. Organizations with well-designed and executed mentoring efforts are key strategies for developing a diverse pipeline of leaders. The impact of peer-to-peer mentoring is also a powerful tool not

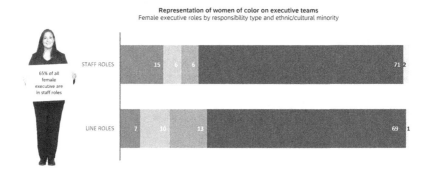

FIGURE 5.2   Representation of women of colorcolor on executive teamssteams.

Graph compiled using data from McKinsey & Company, Diversity Matters.

only for developing leaders, but also in shaping an organization's culture in a way that values diversity and inclusion (Giscombe, Mattis, 2002).

BOARD OF DIRECTORS

Mid-size and large companies typically have a Board of Directors, which is a group of members affiliated with the company in some way who are responsible for overseeing the large-scale activities of the company. Choosing a Board of Directors is an important step in creating the business structure. Typically, executives, and sometimes even the full organization, vote to decide who will be members of the company's board. In companies that have gone public on the stock market, these directors are chosen by the company's shareholders. Some types of corporations require you to have a Board of Directors by law. The typical duties of Board of Directors are explained in the below section.

RESPONSIBILITIES OF THE BOARD OF DIRECTORS

Before choosing a Board of Directors, it is important to know what the board's responsibilities are. While the executive management team like the president, CEO, and CFO are all there to handle the day-to-day responsibilities, the board often includes outside experts to help to generally guide the company. This is specifically important when a company is deciding how to implement a diversity initiative. Often, the board will have significant input in these matters. Other duties of a Board of Directors are usually outlined in the company bylaws. Generally, they include the following:

- Shaping the Company Mission: A Board of Directors often has the ultimate input on the direction and values of the company. This often culminates in the drafting or refining of a company mission. This mission states the overall goals and values of the company and lets the public know what the company's priorities are. When it comes to prioritizing diversity, this is an important area where a Board of Directors should consider including language regarding a company's diversity initiatives.

- Approval of Executives: In many large companies, the Board of Directors will have final say in the appointment of executive roles, including the C-Suite and other top management positions. This responsibility presents an opportunity for the board to assess whether a diverse pool of applicants is being considered for these positions and to emphasize diversity when appointing these top-level positions.

- Financial Advisory: Oftentimes, the board is responsible for advising company financial policy. With diversity's impact on increasing company performance, it is important for board members to be aware of these impacts and work to make sure diversity is implemented for the company's benefit.

DIVERSIFYING THE BOARD

For a company that values diversity, it is important that the makeup of its Board of Directors reflects the diverse makeup of a company's employees or sets an example for what diversity in other parts of the company should look like. If this is the goal, then companies must use unconventional metrics for evaluating what a valuable board member looks like. In the past, companies have used a standard set of C-Suite credentials such as decades of business experience or long-term leadership positions to evaluate whether a candidate will be a valuable board member. However, looking at candidates who may have different kinds of experience, such as unconventional training in a field other than business, can help strengthen the board while also broadening its diversity (Liu, 2022).

In the past, C-Suite and board members typically got their positions through strong and extensive personal networks. Networks such as these have historically left out underrepresented groups like minorities and women. To broaden the pool of candidates for these important roles, it is important to look beyond an immediate circle of well-connected people when searching for the best candidates. One way to bring a fresh set of eyes to this process can be to bring in an outside search team, which can help find diverse candidates both inside and outside of these networks. Traditionally, companies make up their boards almost entirely of CEOs and CFOs, but adding different candidates with varied experience can contribute diversity of thought and valuable outsider knowledge to the board. Having board members with experience as industry experts or with customer interfacing can provide a broader perspective that can make the company as a whole more in-tune with the people it serves, and more successful as a result.

Many organizations that advocate for minority professionals have learned how to work with those selection committees to advance toward board diversity. For instance, the Black Young Professional Network works with companies like Google, Facebook, and Adobe to connect its members to leadership roles in these companies.

IMPORTANCE OF A DIVERSE BOARD

Diversifying a company's board allows for the contributions of a wide range of different perspectives brought by different life experiences such as gender, class, ethnicity, or age. This can influence and improve decision making, help when performing advisory duties, and provide varying perspectives on important topics like risk management. While diverse board members may not have years of experience as a CEO, they will have lived and experienced many other trials and challenges that impressed the board selectors, and these attributes can be just as valuable if not more, than the traditional experience sought for board members.

BOARD DIVERSITY POLICY

One way to set a path for achieving board diversity is to implement a board diversity policy. Setting written expectations for a diverse board provides an avenue for the public, investors, and other stakeholders to hold the board accountable for implementing more diverse selection practices.

RELEASING A BOARD DIVERSITY REPORT

Another way to hold the board accountable for implementing a diversity policy is to publish a periodical diversity report. These reports typically include several metrics, observations, and milestones to inform stakeholders of the efforts being made to achieve strong board diversity. This can include measures such as regularly updating the board's diversity policy as well as the diversity initiative for the entire company, setting diversity progress markers and noting whether they were met, detailing resources such as training or diversity events that were held by the company, and forming partnerships with any organizations that helped advance the diversity of the company and the board.

CEO PLEDGE

One example of a collective effort to achieve diversity and inclusion at the executive level is the CEO Action Pledge for Diversity and Inclusion. Formed by several CEOs from major companies in 2017, the group wanted to tackle the issue of addressing diversity and inclusion in the workplace, starting with those who lead them. The group recruited over 350 companies from across the world to join in the pledge. Their aims are to achieve diversity and inclusion progress among their companies by all working together. The pledge specifically sets a plan of action that the members have agreed to, helping create welcoming and inclusive environments at their companies.

The CEOs who take The Pledge for Diversity and Inclusion commit four initial goals that will catalyze further conversation and action around diversity and inclusion within the workplace and foster collaboration among the organizations:

1. Make workplaces open and trusting places to have complex, and sometimes difficult, conversations about diversity and inclusion

2. Implement and expand unconscious bias education, which will enable individuals to begin recognizing, acknowledging, and therefore minimizing any potential blind spots they might have, but were not aware of previously. This will help the employees recognize and minimize their blind spots and aim to facilitate more open and honest conversations.

3. Share best practices as well as unsuccessful ones.

4. Create and share strategic inclusion and diversity plans with the Board of Directors. The Board of Directors (or equivalent governing bodies) work through the development and evaluation of concrete, strategic action plans to prioritize and drive accountability around diversity and inclusion.

CHIEF DIVERSITY OFFICER

Another way a company can champion diversity and inclusion in the C-Suite is by appointing a Chief Diversity Officer (CDO). The main responsibility of a CDO is to ensure an open and inclusive workplace is cultivated and to champion programs that support the company's diversity and inclusion initiatives. They are often tasked with spearheading programs such as diversity trainings and events that celebrate diversity and culture. CDOs are also the main advocate for ensuring that the company is inclusive for all and that all employees, regardless of any minority status, feel comfortable in the workplace. It is the CDO's responsibility to monitor at a high level this aspect of company operations, identify any shortcomings the company has in terms of diversity and multiculturalism, and be involved in investigations of discrimination or harassment.

HOW DIVERSITY OFFICERS CHANGE CORPORATE CULTURE

As companies continue to take workplace diversity, inclusion, and equity more seriously, CDOs, who are charged with creating policies and climates supportive of workers from an array of backgrounds, have become a more common presence at most companies. The presence of a diversity executive in the C-Suite is one metric for assessing whether a company is equipped to hire and retain diverse workers and effectively market to the heterogeneous customer base of the future.

The work of diversity officers spans across several departments and levels of a company. As discussed in the above section, diversity officers are responsible for shifting the company culture to one of appreciation for diversity and for the creation of a welcoming and inclusive environment. They are also responsible for ensuring equity, or fairness in all company practices. Federal policies such as Title VII of the Civil Rights Act, which bans discrimination in the workplace on the basis of sex, race, color, religion, or national origin, and the Americans with Disabilities Act, which requires that the workplace provide accommodations for those with disabilities, ensure that Americans of all backgrounds have equal and equitable opportunities and rights within the workplace. The CDO must be extremely knowledgeable of these policies and must be aware of how to implement them in their company.

When the company is going through a hiring process, the CDO, human resources, and recruiting teams should work together to remove biases or barriers that would advantage or disadvantage certain individuals based on gender, race, or other characteristics. They may oversee training programs related to unconscious bias, run workshops about communicating effectively in teams, or plan classes about civility in the workplace. To better understand the state of a company's climate, officers may issue surveys asking employees how satisfied they are in their roles, then look for patterns in the results that suggest racial, gender, or other disparities.

The other conditions critical to diversity officer success are reporting directly to the CEO, having a mandate to set strategies, and possessing the authority to hold managers and workers accountable for meeting goals. One of the most important conditions is building a strong business case to justify their work.

CONCLUSION

Research has shown that a company's diversity policy will not be effective unless every level of the company is engaged in the mission of diversity and inclusion. This is especially true of a company's C-Suite. These executives are responsible for setting the mission and high-level goals of the company. Without their active support and engagement in these initiatives, it is unlikely that a strong inclusive culture will be able to take off at other levels throughout the company. The C-Suite is responsible for allocating the funds that can support these programs and spearheading initiatives that will make employees want to change the workplace culture. They are also the final say in how the company is managed and what inclusive practices will ultimately look like. Working to make the C-Suite as diverse as possible and including specialized members such as a Chief Diversity Officer will set the company on the right path for having a truly effective workplace of inclusion.

Chapter 5 Review Questions

1. What constitutes a C-Crew in a corporation?

2. What are the experience, qualifications, and the duties of C-Crew members?

3. Discuss the relationship between innovation and increased diversity in a corporation.

4. Is holding managers and the executives accountable for diversity objectives helping productivity?

5. Describe how senior management is likely to achieve greater profit by inducting gender diversity in their team.

6. Does representation of women of color in the executive team benefit the organization?

7. Discuss the roles of "Board of Directors," and what is a diversity board?

8. Discuss the board diversity policy and details of its reporting method.

9. Why do some corporations have an action pledge for diversity and inclusion?

10. What is the benefit of having a Chief of Diversity Officer (CDO)?

REFERENCES

Barrington L. (2001). Workforce Diversity and Productivity: An Analysis of Employer-Employee Match Data.

Diversity wins: How inclusion matters. (2020, May 19). McKinsey & Company. https://www.mckinsey.com/featured-insights/diversity-and-inclusion/diversity-wins-how-inclusion-matters

Giscombe, K., & Mattis, M. (2002). Leveling the Playing Fieldplaying field for Womenwomen of Colorcolor in Corporate Management:corporate management: Is the Business Case Enough?business case enough?. *Journal of Business Ethics - J BUS ETHICS, 37*, 103–119. 10.1023/A:1014786313354

Liu I., (2022, November 9). *Walking the walk: How companies can spur true board diversity.* Thomson Reuters Institute. https://www.thomsonreuters.com/en-us/posts/news-and-media/corporate-board-diversity/

Economic Benefits from Diversity

E conomic benefits are the benefits that can be quantified in terms of income, profits, revenues, etc. Diversity drives economic benefits for companies. A diverse workforce is essential to a strong economy. Businesses that embrace diversity have a more solid footing in the marketplace than others. A diverse workforce combines workers from different backgrounds and experiences that together breed a more creative, innovative, and productive workforce. In addition, businesses have learned that they can draw upon our nation's diversity to strengthen their bottom line. In this way, diversity is a key ingredient to growing a strong and inclusive economy that is built to last.

TOP ECONOMIC BENEFITS

Driving Economic Growth

One of the top benefits of a diverse workforce is that it helps drive economic growth. As the workforce becomes more open and accepting of people from all walks of life, human capital in the economy increases. A study by Mckinsey and company found that the increase in women in the workforce over the past 40 years has accounted for approximately 25% of the current GDP in the United States (Dixon-Fyle et al., 2020). The American Sociological Association also found in a study that the most diverse businesses made nearly 15 times the revenue of the least diverse businesses (Science Daily, 2009). Allowing more workers into the workforce, increases beneficial economic factors like minimal worker shortages and more efficiency, effectively increasing productivity and invigorating the economy as a whole.

DOI: 10.1201/9781003340966-6

Better Connection with Consumers

As discussed throughout this book, having a diverse team of employees helps companies connect better with their customers. Employees from minority groups will understand firsthand how to best connect with the consumers with similar backgrounds. Employees from younger generations can also provide valuable insights on up-and-coming trends due to use of non-traditional channels such as social media. They can provide valuable insights from their lived experience that a homogenous team would not have. This can be especially beneficial when testing products or marketing strategies, potentially saving companies significant amounts of time and money in failed ad campaigns or unpopular products. It has even been found that companies with strong diversity were 70% more likely to make gains in new and emerging markets than companies that lack diversity.

Better Qualified Workforce

While the need for diverse hiring and recruiting practices has to be highlighted in previous chapters, there can also be a serious economic advantage of implementing these hiring practices. Ensuring candidates are recruited from the most diverse pool possible increases the success with which qualified candidates can be found. By broadening the group of candidates a company is trying to recruit, there is a higher likelihood that a candidate can be found who meets or exceeds the needs of the company. Recruiting qualified candidates can be an extreme economic benefit and avoids losses experienced due to employee job training or excessive turnover.

Avoiding Employee Turnover

Excessive or consistent employee turnover can be extremely detrimental to a company's profits. Losing several employees in short periods of time can cause decreased productivity, knowledge gaps, and unhappiness among employees who remain as they will have to fill in the gaps left by past employees. While many factors can contribute to turnover, ensuring a welcoming environment, especially for diverse employees, can help to keep these workers satisfied with their work life and more likely to remain at the company for a longer period of time. Conversely, if businesses create or allow a hostile work environment for its most marginalized employees, this will create an incentive for these employees to leave the company quickly. Not only does an inclusive workplace encourage employees to stay with a company, but businesses that invested in fostering a work environment of inclusivity saw more than double the cash flow per employee over a period of three years.

THE BUYING POWER OF DIVERSE CONSUMER MARKETS

Businesses are increasingly benefiting from diversity as consumer markets become more diverse and those diverse swaths of the market increase their buying power. Buying power is the ability of consumers to purchase goods or services. Typically, buying power increases with income. The buying power of people of color is approximately $1.2 trillion, and companies that have appealed to these consumers have seen increased profits. The buying power of Hispanic consumers is $1.3 trillion. Focus groups have shown that these consumers are more likely to reward companies that pay attention to their needs as a group, so businesses that make good faith efforts to reach out and make products for these communities can create a strong relationship and consumer base. The LGBTQ community's buying power is over $900 billion, and it has been shown that companies that value diversity are extremely important to this group. One survey found that 90% of LGBTQ individuals consider a company's treatment of LGBTQ members and sponsorship of LGBTQ when searching for jobs. Three-quarters of LGBTQ adults have also stated that they would consciously choose LGBTQ-friendly brands if they had the option (Ahmed, 2021).

Buying power also varies by generation, as those later in their lives with longer working experience tend to have more money available to spend on consumer items. The buying power of individuals born between 1965 and 1980, Generation X, is approximately $2.4 trillion, while the buying power of the younger generation, millennials, is only about $1.4 trillion (Insider Intelligence, 2021).

As the United States has grown more diverse, so has its buying power. Over the past 30 years, the buying power of consumers who identify as African American, Asian American, or Native American has increased dramatically. Their buying power was estimated as $458 billion in 1990 and was found to be approximately $3 trillion in 2020. These groups' share of American buying power has also increased from only 10.6% in 1990. It is now approximately 17%. The same trend was noted in Hispanic Americans, whose buying power increased from 5% in 1990 to 11% in 2020. Figure 6.1 illustrate this trend of increased buying power among minority groups and show that further growth is expected in the coming years. Americans from

US Buying Power *, by Race/ Ethnicity, 2000-2025

Table 6.1 billions

Race/ Year	2000	2010	2020	2025
White	$ 6,425.00	$ 9,479.00	$ 14,191.00	$ 17,350.00
Black	$ 611.00	$ 979.00	$ 1,574.00	$ 1,978.00
Asian	$ 278.00	$ 614.00	$ 1,297.00	$ 1,802.00
Multiracial	$ 60.60	$ 149.20	$ 286.40	$ 396.80
Native American	$ 40.00	$ 84.00	$ 140.00	$ 178.00
Total	$ 7,415.00	$ 11,306.00	$ 17,489.00	$ 21,705.00

FIGURE 6.1   Past and projected U.S. consumer buying power by race/ethnicity.

Graph compiled using data from Insider Intelligence, 2023.

minority groups have significantly more buying power than in the past, and companies' outreach strategies, product development, and advertising must be altered to reflect this.

PROFIT ANALYSIS BASED ON DIVERSITY

Some of the largest companies in the world have seen the impact of diversity on their profits. We looked into five of these large companies – Apple, CitiGroup, Google, Microsoft, and Walmart – to examine how diversity has had an effect on profits (Table 6.1).

1. **Apple**

The case of Apple's profits during this time period provides a promising argument for the effects of diversity on earnings. In the two highest years of profits for the company, 2017 and 2018, diversity was also steadily rising at the company. During these years, its percentage of White employees dropped from 56% to 53% in 2017 and 50% in 2018. In conjunction with this decrease, Asian and Black employees became a larger

TABLE 6.1   Overview of Data Reviewed for Profit Analysis Study

Serial No.	Company	Data Availability
1	Apple	2014–2018
2	CitiGroup	2014–2018
3	Google	2014–2019
4	Microsoft	2015–2018
5	Walmart	2013–2017

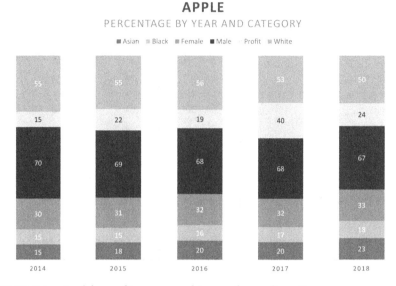

FIGURE 6.2   Apple's profits compared to employee diversity.

share of the company makeup. The number of female employees also slightly increased during this time of increased profits (Figure 6.2).

2. CitiGroup

The demographics at CitiGroup are an interesting case. The only demographic group which significantly increased, from 12% to 18%, were those who identified as Asian. The percentage of both Black and White employees both decreased over this time. The percentage of males and females remained almost stagnant. Aside from an outlying year for profits in 2017, earnings for the company do not appear to be impacted by any of these demographic shifts (Figure 6.3).

3. Google

Google saw significant increases in diverse groups between 2014 and 2019, with a 10% increase of Asian employees and a 3% increase in female employees. Between 2014 and 2018, the company saw increasing profits with exceptionally high profits in 2017 and a decrease in 2019. Even with these variations, between 2014 and 2019, profits generally tend to increase as the company diversified (Figure 6.4).

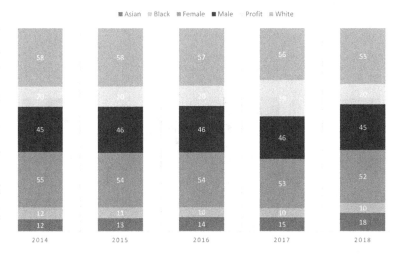

FIGURE 6.3   CitiGroup's profits compared to employee diversity.

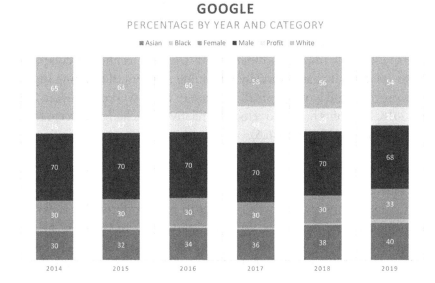

FIGURE 6.4   Google's profits compared to employee diversity.

4. Microsoft

Microsoft also saw significant profit increases between 2015 and 2018. During this period, the number of Asian, Black, and female employees slowly but steadily increased (Figure 6.5).

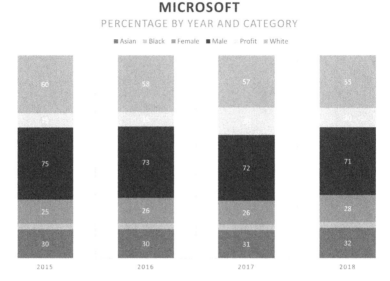

MICROSOFT

PERCENTAGE BY YEAR AND CATEGORY

■ Asian ■ Black ■ Female ■ Male Profit ■ White

FIGURE 6.5   Microsoft's profits compared to employee diversity.

5. ■ **Walmart**

Due to the nature of Walmart's business as a commercial retailer, its diversity statistics and profits differ slightly from the technology companies above. Its diversity has essentially remaining stagnant, fluctuating slightly between years and having a small increase in Black workers between 2014 and 2018. Its profits have also remained relatively unchanged, except for an outlier year of 2017 (Figure 6.6).

BENEFITS IMPACTING THE UNITED STATES

Not only can individual businesses benefit from the impacts of diversity, but the U.S. economy as a whole stands to gain from trends of diversity and inclusion in business. In 2010, the percentage of the GDP held by minority-owned businesses was only 1.6%. This figure is now almost 3% and is expected to reach over 7% by 2060. Currently, 1.9 million businesses are owned by women of color, and they generate over $165 billion annually in revenue. They also employ over 1 million people. It has also been found that the national GDP would increase by $25 billion if just 1% more disabled people were hired.

One study conducted among 195 countries found that countries experienced higher economic prosperity with higher diversity of immigrants. This was found to be particularly true when skilled workers

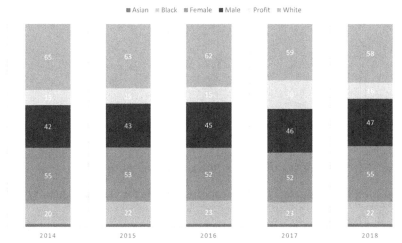

FIGURE 6.6   Walmart's profits compared to employee diversity.

migrated to countries with high incomes. It was found that a 1% increase in diversity from skilled migrant workers increased a country's GDP by 2% (Crane et al., 2021). Another study found that over a period of 50 years, the U.S. economy saw positive growth due to increases in skilled immigrant workers (Docquier et al., 2014). Similar patterns were found in European countries as well.

BENEFITS IMPACTING INDUSTRY

When it comes to Science, Technology, Engineering and Mathematics (STEM) and Industry, the impact of diversifying the workforce can be significant. The need for increase in diversity in industry is two-fold. Both women and ethnic minorities are underrepresented in these fields, and inclusion of these groups has many positive impacts, including economic ones.

A study conducted in Europe found that increasing the number of women in STEM occupations could increase the GDP of the EU by up to 3% by the year 2050 (2019). This would equate to a potential increase of approximately €820 billion. These data are represented in Figure 6.7.

Lessening the gap between genders in STEM can have the most impact if this is addressed at the educational level, as it would positively impact future employment. If more women are encouraged to pursue STEM education, it is estimated that 350,000 more jobs could be filled in the EU

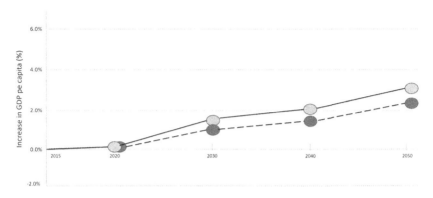

FIGURE 6.7   GDP impact on closing the gender gap in STEM.

Graph compiled using data from European Institute for Gender Equality, 2019.

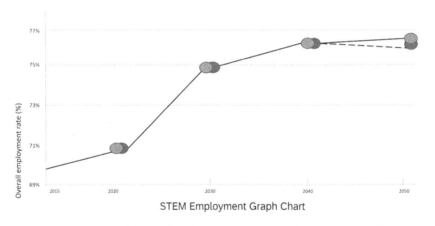

FIGURE 6.8   Impact of closing gender gaps in STEM education on employment.

Graph compiled using data from European Institute for Gender Equality, 2019.

alone (2019). Figure 6.8 shows the impact that closing the gender gap in STEM education could have on future employment figures.

Aside from gender diversity, ethnic diversity contributed by immigrant populations has been found to have a substantial impact on economic benefits, specifically in STEM and Industry.

It is estimated that STEM workers from outside of the United States contributed approximately $400 billion to the U.S. GDP in 2019 alone. This is nearly 2% of the entire GDP for the country.

In addition to contributing to the existing STEM workforce, entrepreneurs from foreign countries have founded research and development companies in the United States, adding up to an additional 1.8% of the U.S. GDP. It was found that the net impact of foreign-born STEM workers in the United States can provide benefits of up to $700,000 per person over only a three-year time period.

THE ECONOMIC IMPERATIVE OF DIVERSITY IN TECH

The positive impact of diversity has shown to be especially prevalent in tech, and the need for even more growth in diversity is imperative in this industry. One noticeable area where improvement is needed is in gender diversity. The number of women in computer science has dropped significantly. In 1985, 37% of computer science bachelor's degrees were earned by women. In 2017, that number was only 19%. Similar trends were found in mathematics. Comparatively, women hold approximately 46% of jobs in the workforce, so the industry as a whole is lacking in gender representation. This problem is compounded when looking at racial ethnic backgrounds – Hispanic, Black, and Asian women hold only 9% of the jobs in tech. Figure 6.9 highlights this disparity among some of the world's leading tech companies (Elberfeld, 2019).

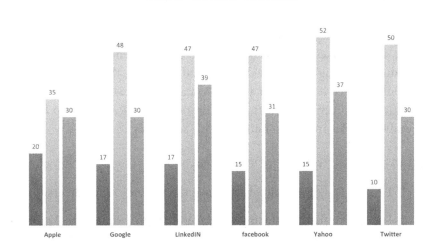

FIGURE 6.9   Gender disparity in major tech companies.

Graph compiled using data from Richter, 2014.

By failing to address the underlying systemic issues causing these disparities, the tech industry is doing itself a disservice in terms of both innovation and profits. A Harvard Business Review study found that companies with high diversity had 19% higher revenues driven by innovation (Prosperix, 2022). The tech industry has a long way to go to achieve higher levels of both ethnic and gender diversity, but they can only stand to gain from making these efforts.

NEGATIVE IMPACTS OF NON-INCLUSION

Not only is diversity and inclusion beneficial to companies for the many reasons previously discussed, but failing to implement inclusion at a company can actually be detrimental. By excluding any minority groups, companies lose out on reaching these populations from both a consumer standpoint and a talent standpoint. One study found that France could increase their GDP by 7% over 20 years if they would work to elevate employment rates of women and disadvantaged groups. Failing to hire from diverse groups also places employers at a disadvantage in the event of a labor shortage. A study in the Netherlands found that shortages of skilled labor positively impacted the hiring of minority groups.

CONCLUSION

Since diversity and inclusion have come to the forefront of public attention, extensive research has shown that there is a strong economic argument to be made for increases in diversity in the workplace. However, it is important to note that companies whose only interest in diversity and inclusion is economic, are not likely to succeed. While economic benefits are a positive byproduct of strong workplace diversity and inclusion, these programs will truly only succeed if those in charge understand the societal need and moral obligation for promoting inclusivity. Diverse workplaces can help lay the groundwork for a more just and equitable society as a whole.

Chapter 6 Review Questions

1. Explain the economic growth and the benefits from diversity.

2. Describe how the consumer connection with a similar background helps the company as a whole.

3. Describe the roles of a better-qualified workforce for extreme economic benefit.

4. Describe the buying power of diverse consumer markets.

5. Examine how diversity has an effect on the profit of large companies.

6. Discuss the benefits impacting the U.S. economy as a whole.

7. Describe the GDP impact on closing the gender gap in STEM and industry.

8. Explain the economic imperative of diversity in the tech industry.

9. Describe the negative impacts of non-inclusion.

10. Explain the gender disparity in major tech companies.

REFERENCES

Alesina, A., Harnoss, J., & Rapoport, H. (2013). Birthplace Diversity and Economic Prosperity. 10.3386/w18699

Ahmed, A. (2021, April 26). What is consumer buying power?. Small Business - Chron.com. https://smallbusiness.chron.com/consumer-buying-power-68682.html

Crane, K.W., Colvin, T.J., Goldman, A.R., Grumbling, E.R., Ware, A.B.. (2021, October). Economic Benefits and Losses from Foreign STEM Talent in the United States. https://www.ida.org/research-and-publications/publications/all/e/ec/economic-benefits-and-losses-from-foreign-stem-talent-in-the-united-states

Dixon-Fyle, S., Dolan, K., Hunt, D. V., & Prince, S. (2020, May 19). *Diversity wins: How inclusion matters.* McKinsey & Company. https://www.mckinsey.com/featured-insights/diversity-and-inclusion/diversity-wins-how-inclusion-matters

Docquier, F., Peri, G., & Ruyssen, I. (2014). The cross-country determinants of potential migration. *International Migration Review*, 48(1_suppl), 37–99. 10.1111/imre.12137

Elberfeld, J. (2019, April 25). *Council post: The economic imperative of diversity in Tech.* Forbes. http://www.forbes.com/sites/forbestechcouncil/2019/04/25/the-economic-imperative-of-diversity-in-tech/?sh=1186c31810d2

How gender equality in STEM education leads to economic growth. European Institute for Gender Equality. (2019, July 23). https://eige.europa.eu/gender-mainstreaming/policy-areas/economic-and-financial-affairs/economic-benefits-gender-equality/stem

Insider Intelligence. (2023, January 31). *US buying power*, by Race/ethnicity, 2010, 2021, & 2026 (Billions).* Insider Intelligence. https://www.insiderintelligence.com/chart/262860/us-buying-power-by-raceethnicity-2010-2021-2026-billions

Prosperix (2022, August 19). *The economic benefits of workforce diversity: Prosperix.* Prosperix | Workforce Innovation Solutions. https://www.crowdstaffing. com/blog/the-economic-benefits-of-workforce-diversity

Richter, F. (2014, August 14). *Infographic: Women vastly underrepresented in Silicon Valley Tech jobs.* Statista Infographics. https://www.statista.com/ chart/2582/female-employment-in-tech-companies/#:~:text=Looking%20at %20tech%20jobs%20(programmers,including%20Apple%2C%20Google %20and%20Facebook

ScienceDaily. (2009, April 3). *Diversity linked to increased sales revenue and profits, more customers.* ScienceDaily. https://www.sciencedaily.com/releases/2009/ 03/090331091252.htm

Development of a Diversity and Inclusion Initiative

Strategy, Implementation and the Roles of Equal Employment Opportunity Law

A big question in diversity management is to ask how can a company increase diversity? We have discussed in the previous chapters that improving diversity levels in the workplace will allow for more of the benefits that come from having a diverse workforce. Competencies that build organizational diversity include communication, cultural self-awareness, knowledge of differences and culture, institutionalization of cultural knowledge, and adaptations to diversity. It must be understood that diversity is about more than race or ethnicity. Recruitment over a broader geographic area, focusing on retention, and paying attention to organizational culture will help promote equality and diversity.

The most essential aspects of improving a company's diversity are treating all employees fairly, creating an inclusive culture for all, and ensuring equal access to opportunities to enable all to fully participate in the work process, which will enable all staff and students to develop to their full potential.

STARTING A DIVERSITY AND INCLUSION INITIATIVE

Employers use diversity and inclusion initiatives for both compliance obligations and to increase the overall bottom line with a more diverse

DOI: 10.1201/9781003340966-7

workforce. Developing a diversity and inclusion initiative involves four main phases:

1. Data collection and analysis to determine the need for change

2. Indentifying Needs and/or Areas of Congern

3. Making Tangible Changes

4. Diversity in Business Objectives

5. Earning Employee Support

6. Implementing Diversity Programs

7. Communicating Diversity Goals and Outcomes

8. Measuring Results and Outcomes

9. Adjusting to Feedback

The following nine steps break down these main phases into action steps employers can take to develop a diversity and inclusion initiative.

Step 1: Collecting Data

The first step to implementing these initiatives is to understand the current state of a company's diversity. This can only be determined if employers understand the current state of the labor market as a whole. They can then compare their company's diversity to that of the full labor market to see where shortcomings may lie. Researching this information can include looking through federal and state labor statistics as well as diversity self-reporting done by private companies. Information to determine the current state of a company's diversity can include protected categories like race, ethnicity, or gender, but it can also include factors like personality types or learning styles. Another good way to understand a company's baseline is to start collecting data on the company itself over time.

Resources to collect necessary diversity data are likely already available at the company. Human Resources departments likely have this data as part of their reporting under Equal Employment Opportunity (EEO) mandates. There is some information, however, that may not be readily available, such as employees' religious affiliations or sexual orientation. This would require voluntary reporting from employees and may be data that are harder to obtain. It is important to be cognizant of the fact that many employees may be skeptical of providing this type of personal information. Because of this, it may be beneficial to use a third-party company or aggregating software that would make the collection of this data completely anonymous, removing employee identifiers and collecting only the data necessary for diversity purposes.

Another data point that can be helpful if conducting a survey of employees is finding their attitudes about the company's current culture surrounding diversity and inclusion. This can reveal whether the company's current diversity tactics are having positive effects or if employees are having any feelings of discrimination that need to be addressed. These surveys can also ask questions about more nontraditional diversity statistics including personality types, learning styles, or life experiences that may impact their working relationships and contribute to diversity of thought.

Step 2: Identify Needs and/or Areas of Concern

After data have been collected as per the methods described above, employers can start to identify groups that are less represented at their company. This can start as a high-level company-wide review and then be narrowed down to understand the demographics within smaller groups like teams, departments, or positions. For instance, this can draw attention to issues such as lack of female or minority representation in management positions, or absence of males in certain departments. It can even help compare differences in diversity among branches of a company.

The information collected from this gathering of data can help to identify areas where the company's diversity efforts may be lacking. It may also be the case that the areas found to be lacking from demographic data do not match employee sentiments of where diversity or inclusion can be improved. Holding focus groups to get a better understanding of these differences in feelings could help create a much better understanding of employees' attitudes toward the company's diversity initiatives.

Step 3: Making Tangible Changes

Once employers have a stronger understanding of the diversity areas that need improvements, it is their responsibility to implement real changes to company policies and practices to help remedy the identified shortcomings.

Changes can include actionable items such as monitoring employee referral programs to ensure that employees are not solely referring candidates similar to them. Unconscious bias, especially in hiring, is another area that a company can tangibly improve upon. For instance, if it is found that a certain department lacks diversity, their candidate selection practices can be reviewed and altered if needed. Another factor that may need to be changed is company culture. If a company has clear preferences toward a religion or political ideology, this may deter candidates or make current employees who do not share those beliefs feel uncomfortable or unwelcome. It is important that a workplace is made to feel inclusive and welcoming to all, so implementing company events that are inclusive to all cultures or beliefs, or ensuring disciplinary action for those who are intolerant of other employees' differences are a few ways to ensure a welcoming and inclusive company culture.

Step 4: Diversity in Business Objectives

The next step to truly incorporating diversity in the workplace is to make sure the company's diversity and inclusion plans match the goals and objectives of the company as a whole. At the topmost level, the company must set diversity-related goals that are both attainable and visible to the community. If a company's goal is to create a diverse and welcoming environment that has engaged and dedicated employees, they should create specific diversity goals like using measurable indicators for diversity such as a diversity index. Other business goals can be indirectly related to diversity, but still require diversity to succeed. If a business wants to increase innovation in its products, it will need to hire diverse thinkers to research and develop products. Having open dialogues for creative decision making where everyone's voice is heard to achieve the best result can only flourish in a welcoming and inclusive environment fostered under the company's direction.

Step 5: Earning Employee Support

As discussed in previous chapters, support from employees at all levels of a company is crucial to the success of a diversity initiative. Support at the senior level is especially important, as those will be the employees setting the tone and example of the initiative and pushing its objectives forward. To this end, it can be beneficial to appoint a senior employee as leader of the company's diversity initiative. This employee would be responsible for outwardly supporting the initiative and keeping the program energized and other employees engaged with it.

One way to help ensure that employees stay engaged with diversity programs is by holding management accountable for ongoing implementation and support of these initiatives. Management should have deliverable tasks such as holding ongoing meetings with staff about creating an inclusive workplace, conducting periodic training on various diversity topics, and enforcing repercussions for employees whose actions do not reflect inclusivity or tolerance.

Another helpful presence in a workplace trying to kickstart inclusivity is an employee committee composed of workers of varying levels throughout the company. This committee would have similar objectives to the senior-level diversity leader, in that they would have to implement the initiatives by engaging their fellow employees in the process. By including employees of all levels on this committee, there is a potential for more employees to be engaged in these efforts as they would be encouraged by their peers. However, a committee such as this can only flourish when given direction, goals, and support from the executive level.

Step 6: Implementing Diversity Programs

Diversity programs or initiatives can take many forms. They can include the shifting of company policies to be more inclusive, the training of employees to understand how to foster inclusivity, changes in recruiting and hiring practices, and many other policies or actions. To achieve the goals of these programs, employers must have a plan of action with tangible deliverables that can measure progress and success. Table 7.1 shows an example of what a strong action plan looks like. Clear goals and timelines are set, and accountability is assigned to various employees (mwah, 2023).

Step 7: Communicating Diversity Goals and Programs

In order for employees to be engaged in diversity initiatives, it is important that the goals of diversity programs are communicated effectively to all employees who would be impacted. Not every employee is the same. Employers must understand this and tailor diversity messages to different groups so that all communications are received and clearly understood, allowing the company's efforts to be fully realized. Various types of communication reach different people, so including multiple forms of communication about diversity initiatives such as newsletters, emails, and other forms of announcement can engage a wider group of people.

TABLE 7.1 Sample Diversity and Inclusion Action Plan

		Diversity and Inclusion Strategy – Example			
Goals	Objective	Actions	Responsible Departments	Timeline	Metric
Workforce	Recruitment of diverse and qualified candidates	1. Broad advertising of positions over a range of media 2. Ensuring diverse candidates represented throughout selection process	1. Human Resources 2. Hiring Managers	First Quarter	Diversity Representation Numbers should increase
Environment	Foster a collaborative, flexible, and fair culture to retain employees and help them advance	1. Hold inclusive Leadership Trainings 2. Form groups for employee advocacy	1. Senior Leadership	End of Year	Measure Participation in leadership training, from at least three employee advocacy groups
Long-Term Sustainability	Identify barriers to inclusion and create a system to break them down. Ensure diversity and inclusion is embedded across company policies	1. Review current company policies and practices to begin removal of systematic inequalities 2. Create and implement a performance indicator of diversity and inclusion	1. Human Resources 2. Senior Leadership	End of Year	Company policies and practices have been reviewed and updated Performance indicator is implemented as an employee evaluation tool

Step 8: Measuring Results and Outcomes

It is critical that diversity initiatives have quantifiable or measurable indicators that can track the progress of a company's programs. Some factors can be easily measured, such as company-wide diversity statistics. These numbers can clearly show whether inclusive hiring practices or targeted recruiting are successful. Other measures may take more effort, but can provide valuable insights. Periodic anonymous surveys to gauge employee attitudes on the success of initiatives can be one way to get constructive feedback on the success of programs. Some measures may not be quantifiable, but are still worth keeping track of, such as employee turnover and company recognitions. Success of the diversity programs themselves can also be quantified by tracking participation in diversity events and trainings as well as the frequency of these events.

A few key diversity metric calculations are as follows.

Ratios

One of the simplest ways to measure diversity is through ratios. It is important to compare the ratio of marginalized to non-marginalized groups at the company. The goal would be to have a ratio close to 1:1. For instance a few ratios that can be measured are:

1. Number of Male Employees: Number of Female Employees

2. Number of White Employees: Number of Non-White Employees

3. Number of Employees Under 40 Years Old: Number of Employees Over 40 Years Old

These ratios will need to be evaluated with nuance, and they should be looked at through various lenses. For example, if a company has an equal ratio White employees to non-White employees, but only White employees are executives or in management positions, there is still a wide disparity at the company that needs to be addressed.

Retention and Turnover Rates

As discussed previously, a low employee turnover rate and strong retention rate can be indicative of successful diversity and inclusion policies and the creation of a workplace where employees feel welcome. Retention rate is examined over a specified period of time. Within this time period, the number of employees who left an organization is compared to the number of employees who were initially at the company in terms of a percentage, as shown in Equation 7.1.

Employee Retention % = (Number of Employees at the end of time period)

/(Number of Employees at the beginning of time period) * 100

(7.1)

This employee retention rate can be used on a wide scale or on a narrow scale, depending on the information needed. The calculation can include or exclude populations as needed. For instance, retention can be calculated just for women employees or for non-White employees. It can also be used to understand the retention of certain groups such as managers or executives.

The 80% Rule

One important rule used as an ongoing metric for diversity is the 80% rule, or the four-fifths rule. This rule states that the rate at which minority groups are selected for advancements, hiring, or promotions should be at least 80% of the rate as White non-minority groups are advanced, hired, or promoted. This can be calculated as shown in Equation 7.2.

Promotion Rate % = (Number of Employees Promoted + Number of Employees Hired)

/Total Number of Applicants   (7.2)

The promotion rate should be calculated for each group being evaluated, including any minority groups and the non-minority group. Once the promotion rates are calculated, the minority group promotion rate is compared to the non-minority group promotion rate. If the minority group's rate is more than 80% of the non-minority group's rate, the company is making good progress toward diversity, although the ultimate goal would be to reach 100% equity in this regard.

Diversity Index

To calculate diversity progress for a single instant in time, a diversity index can be used. The index can calculate diversity of a single team or of the company as a whole. A diversity index is created specific to a company's view of which metrics are important to be included in the index. For the simple example below, the metrics of female employees,

TABLE 7.2 Sample Diversity Index

Diversity Group	Number of Employees in diversity group	Total Number of Employees	Diversity Index (Number of Employees in the Diversity Group/Total Number of Employees)
Female Employees	4	10	40%
Non-White Employees	5	10	50%
Employees Over Age 40	3	10	30%
Total	**12**	**30**	**40%**

non-White employees, and employees over the age of 40 were used in a sample team of ten employees: (Table 7.2)

By evaluating the diversity index, a company can examine their diversity metrics as a whole as well as how their diversity index differs for different populations.

Pay Inequity

The gender pay gap at a company can be used as a key metric for gender diversity. This metric is calculated by finding the median salaries of all female employees and all male employees at a company. Once the median of both salary ranges is determined, they can be compared using a percentage or ratio, as shown in Equation 7.3.

$$\text{Gender Pay Equity \%} = (\text{Median Female Salary/Median Male Salary}) \times 100 \quad (7.3)$$

Ideally, the pay equity percentage should be 100%, indicating equal median salaries for both men and women. The closer a company comes to 100% pay equity, the more successful are its gender diversity initiatives.

In addition to tracking these metrics, it is important to communicate these results to all concerned stakeholders. Alerting employees to the ongoing progress of these initiatives keeps everyone engaged and holds management accountable. Publishing these outcomes can also signal the efforts to prospective employees who may value this information.

Step 9: Adjusting to Feedback

While keeping track of diversity data and progress is important, the biggest reason for tracking these metrics is to learn and adjust the company's diversity strategy accordingly. It is important to put procedures in place that allow for periodic review and adjustment of diversity policies. By using metrics as described in Step 8, it should be clear if a specific policy is causing the opposite of the intended effect in terms of diversity inclusion. These periodic reviews can help identify these scenarios so a plan to change the policy can be initiated.

DIVERSITY IMPLEMENTATION CHALLENGES

Even once a strong diversity initiative has been put into place, there are many challenges that can occur. One of the biggest challenges that most companies face is that diversity, equity, and inclusion is a relatively new subject to be incorporated into business strategies. For companies that have operated for decades or even centuries, old business habits may be hard to break. Roughly 60% of all businesses have underdeveloped diversity initiatives, largely because many companies are still in the learning stage (A, 2022). While many companies realize the value and necessity of diversity, there are still a large number of companies who do not reflect a diverse workforce at the most senior level. As discussed previously, inclusivity at the senior levels of a company is crucial for an effective and successful diversity and inclusion initiative. Once a diversity initiative is underway, it is important to not let it lose steam. While it may be started with strong intentions, several other factors in the workplace may compete with the time that can be dedicated to these initiatives. Even in a survey of diversity and inclusion professionals, more than 40% of respondents concluded that only one-fifth of their time was spent on diversity and inclusion initiatives. It is crucial that these programs are made a priority for their ultimate success.

WHY DIVERSITY PROGRAMS FAIL

While companies across the country and the globe have been starting diversity and inclusion initiatives, many are failing to see tangible outcomes. Much of this has been caused by poor program implementation. In forcing aspects of a diversity program like training and performance reviews without a shift in company culture, social scientists have found that arbitrary rules will make people less likely to comply with diversity programs. A study of more than 800 companies in the United States found that genuinely engaging employees and managers in diversity efforts rather than trying to make mandatory requirements will allow for more diversity in the workplace. Engaging tactics such as targeted recruiting, mentor programs, and dedicated task forces were found to be some of the most successful implementation strategies (Dobbin, 2022).

Tactics typically supported by executives tend to be commanding statements of "dos" and "don'ts" that make enforcement simple from the topmost levels of an organization. However, sociological and psychological research has shown that this does not align with what motivates people to make changes. Top-down efforts such as these can go wrong for a few key reasons:

1. The positive effects of one-time diversity training have been found to last only a few days, as the memory of what was learned in the training can be easily forgotten if not reinforced. As a matter of human psychology, people often respond to mandatory courses with resistance, further lessening the effectiveness of mandatory trainings. Implementing diversity awareness into the everyday workplace instead of one-off trainings, could be one of the strongest ways a company can improve its diversity.

2. Hiring tests to determine the skills of workers without taking into account an applicant's identity has also shown to have unintended consequences. It has been found that bias is still showing through, with management selectively deciding who actually needs to take these tests. Oftentimes, managers will only test outside hires or strangers who are more likely to be minorities, whereas applicants whom they do know or who are from inside the company would not be required to take these tests. If every applicant does take the test, management still has discretion to disregard test results, which can lead to further bias. In fact, companies that have job tests for management positions found that managerial positions held by women and minorities actually decreased between 4% and 10%.

3. Many companies feel that having grievance procedures in place for employees to voice their concerns about diversity issues within the company will help bring to light all inclusion issues and will weed out the bad actors. However, poor implementation of these procedures can actually have an opposite effect. While the intention is to create a safe space for employees to voice concerns within the company, many in management may be unhappy with these complaints and may end up retaliating against employees who complain. In 2015, approximately 90,000 discrimination complaints were brought to the

Equal Employment Opportunity Commission (EEOC). Of those complaints, 45% included mention of retaliation after reporting a grievance to their employer. This also perpetuates a cycle that if employees see that grievances are not being taken seriously, they will be less likely to file complaints. With fewer complaints, employers will be less likely to recognize that there is even a problem. These cycles can be caused by rigid grievance systems. If more flexible systems are put in place that focus more on reconciliation or mediation rather than outright punishment, retaliation may be avoided.

While we have discussed common failures of diversity programs, there are many approaches that go against these traditional diversity programs and can achieve higher success. One main commonality to diversity programs that have seen successful is focusing on engagement and experience rather than control. These programs tend to be framed more positively from the start. By tapping into individuals' personalities and desires, companies can find ways to get true engagement in their diversity initiatives. Most people enjoy the company of others and engaging with their colleagues in a non-traditional environment. Many people also have a desire to come across well to those that they socialize. These key ideas about people can give insight into the types of diversity programs that succeed.

Some of the most successful diversity initiatives have implemented programs that speak to peoples' social needs, such as voluntary training. By electing to participate in diversity trainings, the employees do not feel a force from their employer, but rather feel they are taking actions to better themselves. Another program that plays to this idea is the concept of a self-managed team. In these teams, everyone is on an equal playing field working together to reach common goals. Without an authority figure such as a manager, the team is more likely to be open to all others' ideas without the power dynamics and imbalance that often comes with a managerial structure. This type of team building can allow for closer and more accepting working relationships among coworkers.

Another program that has had extremely positive results is the formation of diversity task forces. These groups are composed of a variety of individuals within a company who are committed to the goal of increasing diversity and inclusion. Implementation of these task forces have had stunning results. In a study of 829 U.S. companies, when task forces were put in place, representation for White women in management increased by 11.6%, representation for Black women increased by 22.7%, and representation for Asian women increased by 24.2%.

College recruitment programs aimed at hiring employees from minority groups have also been shown to be effective in increasing diversity. The idea of recruiting promising employees from a larger-than-normal pool of candidates is a positive idea that can get many managers excited. The same study mentioned above found that representation of White, Black, Hispanic, and Asian-American women in management rose by approximately 10% within five years of implementing college recruitment. It was also found that the share of Black male managers increased by 9%.

ROLES OF EEO IN DIVERSITY INCLUSION

We have discussed at length the benefits of implementing strong diversity and inclusion in the workplace. However, especially in the United States, there are some areas of diversity and inclusion that are not optional and that are governed by law. The primary laws governing diversity and non-discrimination in the workplace are called EEO laws. These are laws at the federal and state level that protect employees from hiring and employment practices that may discriminate against or harass individuals. EEO covers all types of identity-based discrimination, including race, gender, age, national origin, religion, disability, pregnancy, and several other categories. For a complete list of protected identities, it is important to consult with local EEOC offices. To ensure a company is adhering to all the appropriate laws, it is advisable that an attorney is consulted to help develop the proper policies in the workplace that comply with EEO laws.

Human Resources

There are several practices that workplaces can employ to ensure their business is complying with EEO standards. The best place to start is with Human Resources (HR). HR is the first line of defense when it comes to EEO compliance, so it is imperative that they are trained on the laws and held accountable for enforcing and practicing them. Promoting inclusivity and openness is not just a great way to foster diversity and inclusion, but it is also a way to ensure EEO laws are being upheld. It is also important for HR to provide open communication channels for any EEO disputes that may arise. By creating open communication, disputes can be more quickly resolved and less often escalated. Resolving disputes early and employing the option of Alternative Dispute Resolution (ADR) can be beneficial to avoiding escalation.

Hiring and Promotion

EEO comes into play significantly during hiring and promotion. It is imperative that recruitment and hiring of new employees be unbiased. It is also necessary to ensure that the pool of candidates is diversified. This is true for new positions as well as opening within the company and promotions. The first step in ensuring the current hiring practices are fair is to analyze the existing hiring process. Identify any practices that could potentially disadvantage minority groups. Preparing a list of exact qualities that are objectively related to the position can also help reduce bias in hiring. By making a specific list and finding candidates who best match these qualities, their personal traits unrelated to job function can be taken out of the equation. It is important that this practice is applied consistently during the hiring process. When coming up with selection criteria, however, it is also important to ensure that none of the criteria selected can inadvertently disadvantage minority groups. For instance, educational requirements should only be included in job postings if a certain education level is essential to the job function, as these requirements can discriminate against certain minority economic or ethnic groups. Something like this may even be a violation of EEO laws.

Employment Conditions

EEO laws are not only in effect when hiring or recruiting, but they also govern the conditions of employment. One major area where this can come into play is compensation. A major aspect of EEO is ensuring that all employees are paid equally for the same work. Ensuring consistency in pay is imperative to fostering diversity and inclusion. When it comes to opportunities for skill development, training, or other experiences, this must also be equally offered to all employees to avoid bias. It is also crucial to make employees feel safe and that they will not be retaliated against should they feel the need to file a complaint. As discussed previously, an open and comfortable line of communication to discuss EEO topics is crucial.

Anti-harassment

While it is important to ensure EEO is being abided by in everyday workplace practices, a strong network to avoid and eliminate harassment needs to be present in the workplace. This can start with a strong anti-harassment policy and must be carried throughout all workplace practices. Important points of an anti-harassment policy include:

1. Clear explanations and examples of conduct that will not be tolerated

2. Reassurance that retaliation for employees who complain will not be tolerated

3. A concise complaint process that allows for various comfortable options for filing a complaint

4. Protection of confidentiality for those who choose to file a complaint

5. A timely complaint process that is taken seriously

6. Clear outline of repercussions and how they will be enforced

COMPLYING WITH EEO

Above, we described a general outline of how a company can implement EEO best practices, but there are some concrete procedures that are compulsory as part of these laws. The first and most obvious obligation is for the company to avoid any form of discrimination. It is also important that the employees know their rights under EEO law. Appendix A shows an example of a poster that should be posted informing employees of their rights (https://www.dol.gov/agencies/ofccp/posters) ("*Equal Employment Opportunity Posters*").

Certain paperwork must also be kept on file for a period of time. Personnel and employment paperwork must be retained for one year by private employers and two years by educational and government institutions. Payroll records should be kept for three years. Any documents related to a discrimination charge must be kept until the dispute can no longer be legally brought to court or until any litigation is fully resolved.

If EEO laws are not followed by an employer, there are several repercussions that can be faced based on the severity of the violation. Small violations for failure to post the "EEO is the Law" poster face a penalty of $525. Companies with 15–100 employees can face EEO claims of up to $50,000, while companies with over 500 employees can face a maximum claim of $300,000.

EEO COMPLAINT PROCESS

If an EEO complaint is filed, there is a standard process that must be followed. Individuals who believe they are the target of discrimination have 180 days to file a charge. This does not need to be submitted by the individual themselves, in the case that they want to remain anonymous.

When a charge is filed, an EEOC investigator will be assigned to the case. It is likely that they will conduct interviews, take statements, and investigate documentation. An employer can also decide to enter into mediation as an alternative. When the investigation is complete, the investigator can either dismiss the claim or move forward with the charge. If the investigation moves forward, EEOC will notify the employer, and all responsible parties will be charged with remedying the situation.

EEO LAWSUIT CASE STUDIES

While companies should strive to implement diversity for the benefits it brings, there are also serious ramifications for companies that do not, at minimum, adhere to laws governing diversity in the workplace. The below case studies will highlight some of the most prominent or influential EEOC cases.

OFCCP v. Bank of America

One consequential and decades-long lawsuit against Bank of America helped to highlight racial discrimination in hiring practices at one of the country's top financial institutions. Two separate instances of hiring discrimination were identified at the company between 1993 and 2005. It was found that unfair hiring practices disqualified Black candidates from entry-level jobs. As a result, Bank of America had to pay over $2 million to the candidates who were affected by the discrimination (Smith, 2019).

EEOC v. Walmart

A 2020 lawsuit against retail giant Walmart brought both gender and racial discrimination to the forefront of the conversation. An employee at one of the retailer's stores in Ottumwa, Iowa, in conjunction with the EEOC filed the suit against Walmart because she was not selected for a management position due to her having a newborn child. The employer assumed that this would cause her to be leaving the company soon. This assumption was considered gender discrimination as the employee did not indicate that she would be leaving her position, and the promotion decision was made based on the employee's motherhood status.

There was also a second element to this case. The employee was a breastfeeding mother who required lactation accommodations during the work day. A White employee in a similar circumstance received a clean office space as her accommodation. Conversely, the employee who brought the lawsuit, a Black woman, was allowed to use an unsanitary storage closet for the same purposes (Segal, 2022).

Abdallah v. Coca-Cola

In 1999, a class action lawsuit revealed widespread systemic discrimination at the Coca-Cola company. The lawsuit was filed accusing Coca-Cola of having a company structure that systematically grouped Black workers at the bottom of their pay scale. Black workers at the company were averaging pay of $26,000 per year less than their White counterparts. It was found that the entire hierarchical structure at the company severely disadvantaged Black employees. In the end, Coca-Cola reached a settlement of $156 million to the affected employees. The settlement also mandated broad changes to the company structure, which cost an additional $36 million. Another unusual condition allowed monitoring of diversity practices by a panel of outsiders (Winter, 2000).

U.S. Dept. of Labor v. Worley Group

The engineering industry has also seen its fair share of discrimination lawsuits. A Houston location of Worley Group found that the engineering and technical services firm discriminated against female, Black, and Hispanic employees during a routine compliance evaluation in 2016. The Office of Federal Contract Compliance Programs (OFCCP) found that female, Black, and Hispanic systems engineers, project engineers, quality control, scheduling, and planning employees were paid less than their counterparts. While the company denies these allegations, a settlement awarded the affected employees $500,000 in back pay and another $500,000 in salary adjustments ("*US Department of Labor, Worley Group reach agreement to resolve alleged gender-based pay discrimination in Houston*").

EEOC v. DLS Engineering Associates, LLC

Another engineering firm was charged with pregnancy discrimination and reached a settlement in 2022. DLS Engineering Associates offered an engineering logistics position to a woman in Jacksonville, Florida. After learning that the woman was five months pregnant, the company then rescinded their offer. The resulting settlement required the company to update their employment policies, provide training on pregnancy discrimination, and send reports to EEOC. They were also required to pay the affected person $70,000 in damages (*"DLS Engineering Associates to Pay $70,000 to Settle EEOC Pregnancy Discrimination Lawsuit).*

CONCLUSION

Proper implementation of diversity and inclusion initiatives in the workplace can be challenging. It is important that companies have an implementation plan that they can stick to, to ensure that their initiatives are successful. Diversity managers should be cognizant of the common reasons why these initiatives fail so they can avoid them or correct these mistakes. It is also important for companies to study which diversity programs have had legitimate success in the past and work to implement those instead of common strategies that have been proven to fail. While many aspects of diversity implementation at a company may be voluntary, it is important to know the laws that govern diversity and inclusion in the workplace and the consequences that can result from failure in implementation.

Chapter 7 Review Questions

1. What are the different phases of inclusion initiatives and their action steps employers should take?

2. Describe the importance of the current state of a company's diversity.

3. How does data collected help the EEO mandates?

4. What are tangible changes a company can make to help remedy the shortcomings of diversity?

5. What are the measurable indicators for diversity in an organization?

6. Is earning employee's respect crucial to the success of diversity initiatives?

7. Describe the goal of diversity progress measurement.

8. In diversity management, what are the measurable indicators that can track the progress of companies?

9. Explain the 80% Rule with examples.

10. Why do diversity programs or their progress fail?

11. Discuss various aspects and agendas of EEO in diversity management.

12. How does HR play a role in EEO compliance standards?

13. Do CEO laws help equal pay?

14. Describe the anti-harassment policy within an organization with regard to diversity.

15. Describe the EEO complaint process.

16. Narrate the case study: OFCCP v. Bank of America, EEOC v. Walmart, U.S. Dept. of Labor v. Worley Group, Abdallah v. Coca-Cola, and U.S. Dept. of Labor v. Worley Group.

17. What are the lessons learned from the case studies in Question 16?

REFERENCES

A. (2022, January 28). *8 of the most common challenges in DE&I today*. Affirmity. https://www.affirmity.com/blog/8-most-common-challenges-dei-today/

DLS Engineering Associates to Pay $70,000 to Settle EEOC Pregnancy Discrimination Lawsuit. (n.d.). US EEOC. https://www.eeoc.gov/newsroom/dls-engineering-associates-pay-70000-settle-eeoc-pregnancy-discrimination-lawsuit

Dobbin, F. (2022, December 13). *Why diversity programs fail*. Harvard Business Review. https://hbr.org/2016/07/why-diversity-programs-fail

Equal Employment Opportunity Posters. (n.d.). DOL. https://www.dol.gov/agencies/ofccp/posters

mwah. making work absolutely human. (2023, February 8). *Sample diversity and inclusion strategy – mwah*. Mwah. https://mwah.live/sample-diversity-and-inclusion-strategy

Segal, E. (2022, February 11). *Walmart is sued for gender and race discrimination by EEOC*. Forbes. https://www.forbes.com/sites/edwardsegal/2022/02/11/walmart-is-sued-for-gender-and-race-discrimination-by-eeoc/?sh=5277c82e5614

Smith, P. (2019, September 30). *Bank of America to pay $4.2 million to settle hiring bias claims*. https://news.bloomberglaw.com/daily-labor-report/bank-of-america-to-pay-4-2-million-to-settle-hiring-bias-claims

US Department of Labor, Worley Group reach agreement to resolve alleged gender-based pay discrimination in Houston. (n.d.). DOL. https://www.dol.gov/newsroom/releases/ofccp/ofccp20210629

Winter, G. (2000, November 17). *Coca-Cola settles racial bias case.* The New York Times. https://www.nytimes.com/2000/11/17/business/coca-cola-settles-racial-bias-case.html

Case Studies

CASE STUDY 1: DIVERSITY AT GOOGLE

We have discussed at length diversity implementation, and specifically in industries like engineering and tech. When a company rolls out a diversity initiative, what pace of success is expected. This case looks at Google's diversity growth based on its 2018 diversity report.

The company's 2018 diversity report showed that while diversity is improving, it is happening at a slower rate than expected. The percentage of White employees at Google dropped by 2.4% between 2017 and 2018, to 53.1%. This also meant that small increases in the percentage of employees from other ethnicities were seen as a result. However, this did not result in significant increases. It can be difficult to see major progress in diversity in just one year.

Looking at the trend at Google over four years, between 2014 and 2018, more significant change is noticed. The company's diversity numbers were much different in 2014 – 61.3% of their employees in the United States were White. Over four years, that number dropped by 8.2%. Figure 8.1 shows an overview of Google's 2018 diversity data (2018).

Google's 2018 hiring practices tended to reflect the changes that they hoped to make at the company. As shown in Figure 8.2, of all new hires at Google in 2017, less than a majority were White, which is a contrast to the overall company representation in 2018, discussed above (Hamilton, 2018).

When examining their diversity growth over these time periods, Google found that diversity growth was slower than expected, but they were becoming more diverse as a company.

The 2018 report concluded that to achieve better outcomes in the company's diversity, their approach needed to evolve. In 2019, the company also broadened their diversity criteria to take survey of those who volunteered to identify as LGBTQ+ and those with disabilities. They also

DOI: 10.1201/9781003340966-8

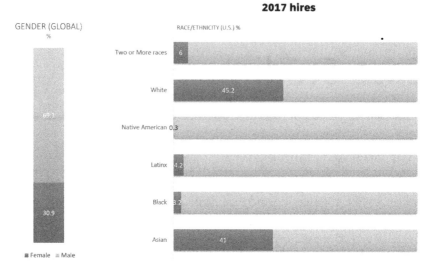

FIGURE 8.1   2018 Diversity data at Google.

Graph compiled using data from Google's 2018 Diversity Report.

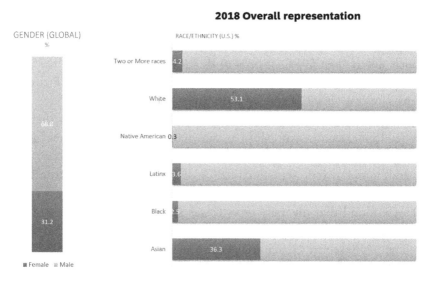

FIGURE 8.2   Google's hiring statistics for 2017.

Graph compiled using data from Google's 2018 Diversity Report.

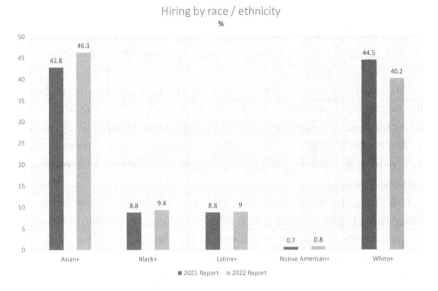

FIGURE 8.3   Google's hiring differences from 2021 to 2022.

Graph compiled using data from Google's 2022 Diversity Report.

took a hard look at many of their inclusion strategies, specifically in hiring and recruiting.

One of the strategies Google highlights in their 2022 report is "meeting people wherever they are" on their career path. Their main emphasis is making a career at Google accessible to people of any background through educational programs for people of all ages. This also included their recruitment strategy, which emphasized equity. This strategy is evident in Figure 8.3. This figure shows the comparison in Google's hiring between 2021 and 2022 (2022). Within that year, their percentage of White hires further decreased by 4%, with significant increases in hires who identify as Asian and smaller increases with member of other minority groups.

This case study shows how a company's diversity policies and initiatives can change and evolve with time. It also shows the importance of not being discouraged by slow progress, as slow progress over time can result in significant change.

CASE STUDY 2: MICROSOFT'S DIVERSITY STRATEGY

The next case study looks at how another successful tech company approaches diversity. Microsoft Corporation decided to test the idea of tying diversity success to compensation. The company decided to try a diversity strategy in which they use potential financial incentives to encourage company-wide participation and enthusiasm for diversity efforts (Statt, 2016).

Microsoft provided several ways for their employees to engage in the company's diversity initiative, all of which would factor into their compensation package. These engagement opportunities included joining a resource group with other employees, taking part in diversity and inclusion training, or simply engaging their peers in productive conversation.

The company took a similar approach to their C-Suite before enacting this company-wide effort. After Microsoft saw steady declines in the percentage of female employees, they decided to take what some saw as a drastic action and tied executive bonuses to the success of diversity initiatives. It was implemented in the hope of creating an incentive for executives to take seriously the need to improve diverse hiring practices.

Since the executive compensation announcement in 2016 and the company-wide financial incentives implemented in 2018, the company has seen some marked improvements. In 2016, the company began these incentives because their percentage of female employees dropped by 1%, from 26.8% to 25.8%. In 2022, their female representation was up to 30.7%. In 2019, after the company was prompted to institute further financial incentives to employees, the company was made up of 33.3% Asian employees, 6.2% Hispanic employees, and 4.4% Black employees. In 2022, these numbers increased to 35.8%, 7.6%, and 6.6%, respectively. Collectively, 53% of the company's employees belong to a racial or ethnic minority (Holman, 2018).

As a result, not only have Microsoft's diversity statistics improved, but the attitudes of their employees toward the company have improved as well. In the company's 2019 diversity report, Microsoft asked if employees felt positively about a sense of belonging at the company. Reflecting the success of their inclusion efforts, 88% of employees responded affirmatively.

Microsoft is an example of a company implementing an incentive-based approach to diversity success. While the monetary incentive is not the company's only diversity strategy, it has shown to contribute to increasing gains in both diversity and feelings of inclusion in the company over several years.

CASE STUDY 3: INFLUENCING DIVERSITY FROM THE TOP

Network Rail is one of Europe's fastest-growing railways. This industrial company of more than 40,000 employees has been striving for more diversity, and this initiative has been strongly led by their CEO and other executives. CEO Mark Carne has stated a belief that diversity will lead to better safety and stronger performance. It also strongly pushes the need to encourage women into male-driven technical fields such as rail, and works to address what they call "inherent bias" in male-dominated fields.

The company set up several initiatives to help achieve these goals, many of them starting from the corporate level. It trained over 2,000 employees in leadership positions in inclusivity and unconscious bias, appointing diversity sponsors at the executive level, tailoring corporate objectives to diversity, and a five-year strategy for the company. On top of these top-level measures, the company dedicated a project team to increasing representation of female employees.

Network Rail also set up a clear hierarchy for their diversity program. Their diversity and inclusion program includes a director, a program manager, and an inclusion manager who are all overseen by a program board that they are expected to report to every quarter. They also have an inclusive leadership program that gives responsibility to managers for ensuring they are fostering an inclusive environment. Leadership in the company has also stressed the implementation of support networks. The goal of these networks is for employees to create strong relationships with their peers with the ultimate aim of creating open dialogues and welcoming environments where people can speak up about diversity and any concerns they may have.

Network Rail has been praised for their diversity efforts and successes. In 2015, they received a silver benchmark achievement from the organization Opportunity Now. This success has not slowed, achieving several recognitions in 2021 and 2022, including the U.K. Times Top 50 Employers for Gender Equality, an award for accommodations for those with disabilities, and a spot in Stonewall's Top 100 for LGBTQ+ inclusion.

We have previously discussed the difficulties of implementing top-down diversity initiatives and the importance of buy-in from management and employees in these strategies. Combining a strong diversity leadership board with a clear reporting structure for lower-level management and an effort to provide an inclusive environment for all employees has come together to spell success at Network Rail.

CASE STUDY 4: IMPLEMENTING LEADERSHIP PROGRAMS

Verizon Wireless, the largest cellular provider in the United States, has seen significant success with diversity programs, specifically their programs for development of women leaders. The company began by identifying a problem they wanted to address and stating a clear plan for implementation of a strategy to address it. Verizon identified that worldwide, there were few women in leadership positions at large companies. They clearly set a goal that elevating women to leadership positions at their company would not be an option, but rather an imperative.

One of the ways that Verizon works toward achieving this goal is through a women's leadership forum implemented throughout the company. The program has the distinct goals of understanding the importance of having women leaders, discussing opportunities to achieve more female leadership, exploring ways to help these women build their careers, and understanding how to retain strong female employees. Verizon's efforts specifically work to help women develop leadership skills through forums, workshops, and other support systems. The program teaches women in the company how they can achieve big successes through a series of small achievements and accomplishments. It also teaches them how to leverage certain traits such as emotional intelligence to work for them and enhance their careers. Additionally, it gives women concrete advice that they can carry into their everyday work life and implement to their benefit.

As a result, Verizon has been recognized both internally and externally for its achievements. The company has been recognized for 17 consecutive years as one of the top companies for multicultural women and in 2021 was in the top 75 companies for women in executive positions. They have also received internal praise from women who have participated in and benefited from these programs.

CASE STUDY 5: GENDER DIVERSITY AT A SMALL ENGINEERING COMPANY

The previous case studies have looked at larger companies in technology and engineering. This case will look at the growth of a smaller engineering firm and how gender diversity has factored into its success. Environmental Engineering Solutions, PC is an environmental and code compliance consulting firm in the New York City Metropolitan area. Founded in 1998, the company has had varying gender makeup throughout its existence.

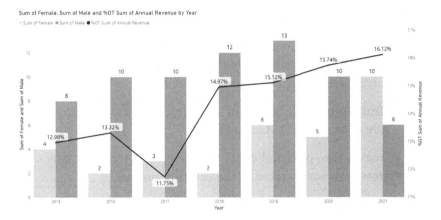

FIGURE 8.4   Profits compared to gender diversity at a small engineering company.

This case study will look at how the company's profits have increased in relation to the company's gender makeup.

Figure 8.4 shows the company's revenue percentage overlaid with gender diversity from the years 2015–2021.

In terms of profits, the company has seen steady increases between 2015 and 2021 with the exception of one year, 2017. Individually, the percentage of male-to-female employees does not significantly align with this trend. However, between the years 2018 and 2021, the percentage of female employees at the company increased drastically, from 14% in 2018 to 63% in 2021. During this same span of time, revenue saw a steady increase in line with this trend. This trend of gender diversity for a small company has been in line with the diversity strategies that larger engineering and tech companies have been championing.

CONCLUSION

Every company can have challenges and triumphs when it comes to diversity programs. Fields such as science, engineering, and technology have especially unique challenges when it comes to issues such as gender diversity and diverse recruiting. Some companies have found unique and successful solutions for addressing these problems. While there is a set of strategies that have often been successful, it is important for a company to be introspective and to carefully examine how diversity can really thrive. It may require trial and error, but the important thing is for companies to keep examining their efforts and persisting until they have found strategies that truly bring success.

Chapter 8 Review Questions

1. Diversity trend in Google for years 2014 through 2018 showed significant change. Provide the cause and the effect. What is the current trend at Google (use current data).

2. Hiring process is crucial in diversity management. Does Google adhere to the EEO standards?

3. Discuss the "pros and cons" with respect to the "company's slow growth versus increase in the diversity process."

4. How does volunteering gender identity affect the diversity, inclusion, and overall harmony in the company's operation?

5. Research indicates that diversity policies and initiatives evolve with time. How can a small organization and STEM industries forecast and implement diversity policies, and reap the prosperity in terms of revenue, growth, etc.?

6. Explain Microsoft's diversity strategies.

7. If a company is adequately staffed by diverse personnel, do employee's attitudes improve towards the company? Provide examples.

8. How, in an organization, is diversity influenced from the top? Provide an example company for this case.

9. How does Verizon Wireless company handle diversity programs?

10. How does gender diversity in a small engineering company affect revenue? (Use EES PC as an example.)

REFERENCES

Diversity annual report – google diversity equity & inclusion. Diversity Annual Report – Google Diversity Equity & Inclusion. (2022).

Diversity annual report – google diversity equity & inclusion. Diversity Annual Report – Google Diversity Equity & Inclusion. (2018). https://diversity.google/static/pdf/Google_Diversity_annual_report_2018.pdf

Hamilton, I. A. (2018, June 18). *These 7 graphs lay bare Google's diversity problem.* Business Insider. https://www.businessinsider.com/google-diversity-problem-in-7-graphs-2018-6

Holman, J. (2018, November 14). *Microsoft ties progress on diversity to how much it pays workers*. Bloomberg.com. https://www.bloomberg.com/news/articles/2018-11-14/microsoft-ties-inclusion-effort-to-workers-compensation-package

Statt, N. (2016a, November 18). *Microsoft says it will tie executive bonuses to diversity hiring goals*. The Verge. https://www.theverge.com/2016/11/18/13681738/microsoft-diversity-goals-executive-bonuses-women-in-tech

Conclusions, Lessons Learned, and Moving Forward

TAKEAWAYS ON DIVERSITY IN INDUSTRY

Diversity and inclusion initiatives in the workplace are constantly changing and evolving. The most successful diversity programs are ones that are adaptable. Over the course of this book, we have looked at the importance of diversity and inclusion, the success it can bring to a company, and the challenges implementing these programs can present. Employing diversity and inclusion in historically homogenous fields such as science, technology, engineering, and industry can bring its own set of unique challenges.

The Definition of Diversity Is Constantly Adapting

Diversity has an ever-expanding and evolving definition, and it can look different at different companies. As the U.S. population becomes more diverse and the world becomes more interconnected, companies need to make a concerted effort for their workforces to reflect the needs of a new population. In addition to racial and ethnic diversity, gender diversity, generational diversity, and diversity of social class all contribute to diversity of thought as a whole. Diversity of thought allows for dynamic and creative teams that allow for innovation. The United States has been slow to grow strong workplace diversity, but has seen some significant improvements in recent years. However, there are still many areas that seems to be overlooked, and specifically in traditionally white male-dominated fields such as science, technology, engineering, and industry.

DOI: 10.1201/9781003340966-9

Throughout the book, we searched for answers on addressing today's most pressing diversity issues, and we found that the need for diversity is both a moral imperative for companies as well as a smart business choice.

Companies Have a Moral Obligation to Implement Diversity, but It Also Makes Economic Sense

There are many economic benefits from increased diversity at both large and small companies. When expanding diversity at a company, having a larger pool of candidates to choose from has proven to create innovative teams with unique ideas that have allowed companies to grow and prosper. Additionally, when employees are brought in from different backgrounds, companies can tap into new markets that align with the diverse communities these employees belong to. Recruiting employees from a large pool leads to finding better-qualified candidates for positions, and creating an inclusive environment helps retain these employees and avoid turnover, cutting the large costs that can come with losing talent and the economic benefits of retaining employees for long periods of time.

Implementation of Diversity Programs Can Be Challenging, but Strong Leadership Can Overcome These Challenges

The need for and benefits of diversity and inclusion are well documented, but the successful implementation of these programs can pose certain challenges. One of the largest challenges is bringing an entire company on board with a diversity strategy and keeping everyone engaged. Diversity and inclusion are truly a team effort, which starts with effective leaders. Achieving diversity in the highest levels of a company, the C-Suite, shows the public a company's commitment to diversity. However, diversity programs often fail if the efforts stay concentrated at the executive level. Incorporating diversity officers into the C-Suite can be valuable, but these officers must understand the importance of outreach and engaging all levels of employees at the company.

Even with Strong Leadership, Diversity Programs Can Only Succeed if All Team Members Are Engaged

The success of a diversity initiative often depends on the choices made by those at the top of the company, but some of the most key players in diversity success work in middle management. While corporate executives may be pushing the company's diversity goals, managers are the individuals who will ultimately determine whether diversity strategies are implemented on a day-to-day basis. A manager who is engaged in the company's diversity program will work to encourage diversity and inclusion among their employees, creating a chain, from corporate down to the everyday employee, of people who are invested in the company's diversity initiatives.

Successful diversity programs come in many forms, but many share some important characteristics. By creating a sense of belonging and inclusion at the company through allowing employees to be themselves and feel welcome in all circumstances, employees will feel valued and safe, leading to positive feelings toward the company and, in turn, their work. To achieve this, leaders, and especially managers, need to lead with empathy, be caring, and try to understand and work with employees when they face issues or struggle. It is also important that companies do not look at diversity in terms of numbers, as quotas can be detrimental to the actual aims of a diversity initiative. The best diversity approach is an ongoing, holistic one where a company truly works with their employees, values their thoughts, and aims to help diverse employees grow as individuals and as leaders.

While the strategies and considerations presented in this book provide a strong foundation for companies looking to further diversity programs for industry, many challenges are unforeseeable. This is especially true in the case of the COVID-19 pandemic. This era presented unprecedented challenges in various aspects of the workplace, and completely altered many diversity initiatives. Next we discuss the importance of adaptability in diversity plans and how companies can use unprecedented scenarios to strengthen and shape their diversity programs.

SUPPORTING DIVERSITY IN UNPRECEDENTED TIMES

The COVID-19 pandemic has changed nearly every aspect of life in some way, and this is especially true when it comes to the workplace. The pandemic posed unique challenges for those diverse members of the workforce, and employers must be aware of these issues in a post-pandemic era. Diverse employees, including women, people of color, and LGBTQ+ employees, have voiced the need for more support during these times. Employers must take these needs into consideration for the well-being of their diverse employees.

The pandemic introduced many new challenges to daily life, most notably the balance between work life and home life. For many, especially women and caregivers, duties at home and with family increased, all while the workplace was being brought into the home. This has led to unique circumstances that employers should be cognizant of and be willing to be flexible in order to support their employees who now face unique challenges at home, such as watching children who have some or all of their schooling remotely or caring for family members and quarantining should someone in their household contract COVID-19.

Diversity in the workforce as a whole was also severely impacted by COVID-19, because the most vulnerable jobs, like foodservice and hospitality, were primarily held by those in minority groups such as ethnic minorities and women. Women with children were particularly impacted, with the average working mother reducing their work hours by 3 or 4 times more than their male counterparts. The burden of childcare plus housework, which disproportionately falls on women and was exacerbated by the COVID-19 pandemic, has caused one in four women to leave the workforce altogether. This is a significant setback for companies, especially in science, engineering, and technology, who are actively seeking more gender diversity.

As we have examined at many points throughout this book, one of the most important factors of thriving diversity in the workplace is a sense of welcoming and a strong camaraderie among employees. The COVID-19 pandemic drastically shifted the workplace dynamic and how workers interact with one another. Roughly 70% of companies implemented remote work during the pandemic, and 62% of those companies plan to continue remote work at their companies to some extent. While flexible hours and remote work were implemented to alleviate the physical and mental health tolls of the pandemic, these policies can dramatically lessen opportunities for in-person interaction and team building with coworkers. All of these factors can present unique challenges for employers.

As a response to these unprecedented disruptions to the traditional workplace, nearly 96% of businesses across the world updated their human resources policies to improve support for employees. Often, this includes changes to support employees' mental health and improvements to diversity and inclusion.

With all the other struggles businesses faced during the pandemic, 29% still ranked diversity and inclusion as a top priority, and 40% ranked it as "very important." Throughout the pandemic, nearly 50% of companies implemented programs to help with personal well-being as well as offering services such as mental health counseling (Dolan et al., 2020). With this reaffirmed commitment to diversity on top of the added challenges of the

COVID-19 pandemic, companies need to update their diversity strategies and add new ones for this unprecedented era.

Consider Everyone's Struggles

The COVID-19 pandemic exposed vulnerabilities that have not been seen in the workforce before. For many employees, this time took a toll on both physical and mental health, and companies learned that to create a truly inclusive environment, they needed to go above and beyond to meet the needs of their employees. Workers' struggles throughout the pandemic also tended to vary based on things like race or class. To keep employees engaged, employers needed to meet employees where they were and provide reasonable accommodations to make their teams feel valued.

Encourage Video Calls

Even with a workforce primarily working from home, there are ways to continue to foster close connections between coworkers to build a sense of community. Encouraging workers to use the same communication channels they would use in person can be a simple way to help keep a sense of community. For instance, if a coworker would normally approach another employee to ask a question, encourage them to continue to have these same kinds of interactions via video call rather than written communication such as an email. Having frequent video calls with cameras on is the closest way to simulate an in-person workplace culture and should be encouraged to keep fostering a strong sense of community in the workplace. It is also important during these calls to make sure standard diversity and inclusion efforts are still being implemented, like encouraging inclusive language and giving space for others to speak up.

Flexibility for Unconventional Situations

More than ever, the COVID-19 pandemic has shown society as a whole the need for adaptability in many areas of life, and especially in the workplace. During the shift to remote work, employees were able to show that productivity can be achieved even if it is not in a conventional setting or even at conventional working hours. The COVID-19 pandemic drastically altered many people's everyday schedules, and in some cases, permanently. Minority employees and women are some of the people most likely to have had these schedule shifts due to childcare and other family responsibilities ("4 ways to make diversity and inclusion the new normal in a post-pandemic era", 2020). One of the best ways employers can help these affected employees feel included and welcome is to have an open mind and be flexible. As long as there are no significant declines in performance, employers should be accepting and understanding of times when their employees may need remote work accommodations or flexible hours. By allowing these types of accommodations, employees will see that their employer truly understands the challenges that can be faced outside the workplace and that there is a sense of compassion that will allow for a comfortable and welcoming workplace. When employees feel welcome and treated well, they feel more empowered to speak up and become more invested in the success of the company.

OPPORTUNITIES

Despite the many challenges of the COVID-19 pandemic, there are several lessons that can be learned to help strengthen companies and their diversity efforts. With diverse groups facing varying inequities due to the pandemic, there are unique opportunities for companies to be at the forefront of shaping diversity policies in this new landscape.

While it can be easy to let priorities like diversity fall to the side during a time of crisis, focusing on diversity can actually be the key to building strength and resiliency at a company. This is not the first crisis in which diversity has proven to help a recovery effort. During the 2008 financial crisis, it was found that the most stable banks amid the crisis were those with larger percentages of women on their boards. These banks may have also been less vulnerable to the impacts of the crisis.

One important task in a time of crisis is to ensure talent is not lost. By monitoring the needs of a company's workforce as well as its demographics, leadership needs to make an effort to be in tune with their employees' desires in order to retain them. Accommodations like remote work and flexible hours will allow companies to retain top talent, and rigidity against these changing dynamics can result in the loss of a company's best minds ("COVID-19 crisis and talent mobility: diversity and inclusion", 2020).

Another opportunity afforded by the pandemic is the chance for important decisions to be made. Because of the unique circumstances of the crisis, unconventional problems will need to be solved. The best companies to solve a diverse new set of problems will be those already positioned with a strong and diverse group of talent. Companies that have an inclusive culture will also have employees who feel empowered to speak their minds and who are not afraid to voice creative ideas.

As we have discussed, diverse teams are often more innovative. In the face of adversity and crisis, innovative and creative teams can be a company's strongest asset. Having a team with individuals from a variety of backgrounds with unique thought processes can help a company navigate through unanticipated circumstances. Studies have found that companies with higher percentages of female employees were more likely to introduce innovative products to the market and companies with diverse leadership were more likely to bring innovative ideas to the market. In the face of an unprecedented pandemic and other unexpected crises, these attributes can be invaluable.

While aspects of the pandemic such as remote work may seem like a drawback to team building and communication, they can actually strengthen a workplace. By allowing flexibility, workers are more likely to find fulfillment and feel appreciated by their employers, causing them to be more receptive to the company's efforts. Times of crisis can also bring people together. It is important to have strong leadership that can make sure a team is brought together by difficult challenges instead of letting relationships erode. Letting employees know they have support during challenging times can also create stronger teams. If possible, compensating employees extra for working through crises will communicate that the company understands socioeconomic challenges brought on by unexpected events, and will allow those impacted, who may be from minority groups, freedom, comfort, and security within the company.

Seizing opportunities such as these in the midst of a crisis may not seem like a priority, but the rewards on the other side of the crisis could potentially be tremendous. Having an existing strong diversity infrastructure can allow the benefits discussed above to come more naturally in unprecedented times.

CONCLUSION

Diversity and inclusion are imperatives in today's workforce, and fields like science, engineering, technology, and industry have a long way to go in catching up with existing diversity standards. Starting and maintaining a diversity program can be a daunting task, but the benefits can be tremendous. A strong diversity and inclusion program can make a company resilient and prosperous. Diverse teams result in higher innovation, and no fields need innovation more than science, engineering, technology, and industry. These industries are at the forefront of today's leading technology and infrastructure, and building a workforce that is just as dynamic through diversity and inclusion is critical toward building a stronger future.

Chapter 9 Review Questions

1. What are the most important takeaways on diversity in industry?

2. Why is the "definition of diversity" constantly adapting?

3. Describe the phenomenon of "moral obligation to implement diversity versus economic sense."

4. Besides the top leaders, all team members in an organization must be engaged for successful diversity inclusion. Tabulate and summarize a few leaders at the leading STEM companies who have accomplished this task successfully.

5. Diversity programs fail if members of the disadvantaged workforce do not step up to take advantage of companies' efforts. Elaborate, with examples and data, to further illustrate this case.

6. How did the COVID-19 pandemic influence the diversity in industry? Which industrial sectors took advantages of this?

7. What were the lessons learned during COVID-19 to strengthen a company, and how did the diversity process contribute to this?

8. How do leaders ensure, in times of crisis, talents are not lost?

9. Are innovations part of the diversity program? Provide an example.

REFERENCES

4 ways to make diversity and inclusion the new normal in a post-pandemic era. (2020, May 20). HR Dive. https://www.hrdive.com/news/4-ways-to-make-diversity-and-inclusion-the-new-normal-in-a-post-pandemic-er/578263/

COVID-19 crisis and talent mobility: diversity and inclusion. (n.d.). Mercer Mobility. https://mobilityexchange.mercer.com/insights/article/covid-19-crisis-and-talent-mobility-diversity-and-inclusion

Dolan, K., Hunt, D. V., Prince, S., & Sancier-Sultan, S. (2020, May 19). *Diversity still matters.* McKinsey & Company. Retrieved April 11, 2023, from https://www.mckinsey.com/featured-insights/diversity-and-inclusion/diversity-still-matters

Appendix

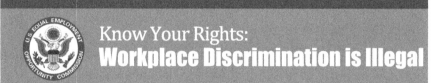

Know Your Rights:
Workplace Discrimination is Illegal

The U.S. Equal Employment Opportunity Commission (EEOC) enforces Federal laws that protect you from discrimination in employment. If you believe you've been discriminated against at work or in applying for a job, the EEOC may be able to help.

Who is Protected?

- Employees (current and former), including managers and temporary employees
- Union members and applicants for membership in a union
- Job applicants

What Organizations are Covered?

- Most private employers
- State and local governments (as employers)
- Educational institutions (as employers)
- Unions
- Staffing agencies

What Types of Employment Discrimination are Illegal?

Under the EEOC's laws, an employer may not discriminate against you, regardless of your immigration status, on the bases of:

- Race
- Color
- Religion
- National origin
- Sex (including pregnancy and related conditions, sexual orientation, or gender identity)
- Age (40 and older)
- Disability
- Genetic information (including employer requests for, or purchase, use, or disclosure of genetic tests, genetic services, or family medical history)
- Retaliation for filing a charge, reasonably opposing discrimination, or participating in a discrimination lawsuit, investigation, or proceeding.

What Employment Practices can be Challenged as Discriminatory?

All aspects of employment, including:

- Discharge, firing, or lay-off
- Harassment (including unwelcome verbal or physical conduct)
- Hiring or promotion
- Assignment
- Pay (unequal wages or compensation)
- Failure to provide reasonable accommodation for a disability or a sincerely-held religious belief, observance or practice
- Benefits
- Job training
- Classification
- Referral
- Obtaining or disclosing genetic information of employees
- Requesting or disclosing medical information of employees
- Conduct that might reasonably discourage someone from opposing discrimination, filing a charge, or participating in an investigation or proceeding.

What can You Do if You Believe Discrimination has Occurred?

Contact the EEOC promptly if you suspect discrimination. Do not delay, because there are strict time limits for filing a charge of discrimination (180 or 300 days, depending on where you live/ work). You can reach the EEOC in any of the following ways:

Submit an inquiry through the EEOC's public portal: https://publicportal.eeoc.gov/Portal/Login.aspx

Call 1–800–669–4000 (toll free)
1–800–669–6820 (TTY)
1–844–234–5122 (ASL video phone)

Visit an EEOC field office (information at www.eeoc.gov/field-office)

E-Mail info@eeoc.gov
Additional information about the EEOC, including information about filing a charge of discrimination, is available at www.eeoc.gov.

EMPLOYERS HOLDING FEDERAL CONTRACTS OR SUBCONTRACTS

The Department of Labor's Office of Federal Contract Compliance Programs (OFCCP) enforces the nondiscrimination and affirmative action commitments of companies doing business with the Federal Government. If you are applying for a job with, or are an employee of, a company with a Federal contract or subcontract, you are protected under Federal law from discrimination on the following bases:

Race, Color, Religion, Sex, Sexual Orientation, Gender Identity, National Origin

Executive Order 11246, as amended, prohibits employment discrimination by Federal contractors based on race, color, religion, sex, sexual orientation, gender identity, or national origin, and requires affirmative action to ensure equality of opportunity in all aspects of employment.

Asking About, Disclosing, or Discussing Pay

Executive Order 11246, as amended, protects applicants and employees of Federal contractors from discrimination based on inquiring about, disclosing, or discussing their compensation or the compensation of other applicants or employees.

Disability

Section 503 of the Rehabilitation Act of 1973, as amended, protects qualified individuals with disabilities from discrimination in hiring, promotion, discharge, pay, fringe benefits, job training, classification, referral, and other aspects of employment by Federal contractors. Disability discrimination includes not making reasonable accommodation to the known physical or mental limitations of an otherwise qualified individual with a disability who is an applicant or employee, barring undue hardship to the employer. Section 503 also requires that Federal contractors take affirmative action to employ and advance in employment qualified individuals with disabilities at all levels of employment, including the executive level.

Protected Veteran Status

The Vietnam Era Veterans' Readjustment Assistance Act of 1974, as amended, 38 U.S.C. 4212, prohibits employment discrimination against, and requires affirmative action to recruit, employ, and advance in employment, disabled veterans, recently separated veterans (i.e., within three years of discharge or release from active duty), active duty wartime or campaign badge veterans, or Armed Forces service medal veterans.

Retaliation

Retaliation is prohibited against a person who files a complaint of discrimination, participates in an OFCCP proceeding, or otherwise opposes discrimination by Federal contractors under these Federal laws.

Any person who believes a contractor has violated its nondiscrimination or affirmative action obligations under OFCCP's authorities should contact immediately:

The Office of Federal Contract Compliance Programs (OFCCP)
U.S. Department of Labor
200 Constitution Avenue, N.W.
Washington, D.C. 20210
1–800–397–6251 (toll-free)

If you are deaf, hard of hearing, or have a speech disability, please dial 7–1–1 to access telecommunications relay services. OFCCP may also be contacted by submitting a question online to OFCCP's Help Desk at https://ofccphelpdesk.dol.gov/s/, or by calling an OFCCP regional or district office, listed in most telephone directories under U.S. Government, Department of Labor and on OFCCP's "Contact Us" webpage at https://www.dol.gov/agencies/ofccp/contact.

PROGRAMS OR ACTIVITIES RECEIVING FEDERAL FINANCIAL ASSISTANCE

Race, Color, National Origin, Sex

In addition to the protections of Title VII of the Civil Rights Act of 1964, as amended, Title VI of the Civil Rights Act of 1964, as amended, prohibits discrimination on the basis of race, color or national origin in programs or activities receiving Federal financial assistance. Employment discrimination is covered by Title VI if the primary objective of the financial assistance is provision of employment, or where employment discrimination causes or may cause discrimination in providing services under such programs. Title IX of the Education Amendments of 1972 prohibits employment discrimination on the basis of sex in educational programs or activities which receive Federal financial assistance.

Individuals with Disabilities

Section 504 of the Rehabilitation Act of 1973, as amended, prohibits employment discrimination on the basis of disability in any program or activity which receives Federal financial assistance. Discrimination is prohibited in all aspects of employment against persons with disabilities who, with or without reasonable accommodation, can perform the essential functions of the job.

If you believe you have been discriminated against in a program of any institution which receives Federal financial assistance, you should immediately contact the Federal agency providing such assistance.

Index

Printed in the United States
by Baker & Taylor Publisher Services